W9-AOS-712

## The Economics of the Middle East

Series Editor: Dr. Nora Ann Colton

The Middle East has seen much more economic change than sociopolitical change over the past few decades in spite of the continuous political instability that is often highlighted by the press. Collectively the region is best known for producing and exporting oil. While the oil industry significantly impacts the region by generating wealth and movement of labor, it has also become the agent of change for endeavors such as development and diversification. With higher rates of growth occurring more in the East than the West, the Middle East sits on the crossroads of this divide acting as a bridge between these two market places. This series is dedicated to highlighting the challenges and opportunities that lie within and around this central region of the global economy. It will be divided into four broad areas: resource management (covering topics such as oil prices and stock markets, history of oil in the region; water; labor migration; remittances in the region), international trade and finance (covering topics such as role of foreign direct investment in the region; Islamic banking; exchange rate and investments), growth and development (covering topics such as social inequities; knowledge creation; growth in emerging markets), and lastly demographic change (covering topics such as population change, women in the labor market, poverty and militancy).

**Dr. Nora Ann Colton** is Principal Lecturer in International Business and Management as well as a Middle East Expert at the Royal Docks Business School, University of East London. Prior to joining the University of East London, Dr. Colton was a Professor of Economics and Business at Drew University as well as the Director of Middle East Studies. Dr. Colton has conducted extensive fieldwork in the Middle East and was a Carnegie Scholar in 2009 and Visiting Professor of Economics at the American University of Beirut.



*Islamic Banking and Finance*
By Omar Masood

*The Global Economic Crisis and Consequences for Development Strategy in Dubai*
Edited by Ali Tawfik Al Sadik and Ibrahim Ahmed Elbadawi

*Land Ownership Inequality and Rural Factor Markets in Turkey*
By Fatma Gül Ünal

*Expats and the Labor Force*
By George Naufal and Ismail Genc

*A Prelude to the Foundation of Political Economy*
By Cyrus Bina

**Other Books By Cyrus Bina**

*The Economics of the Oil Crisis*

*Modern Capitalism and Islamic Ideology in Iran* (coeditor)

*Beyond Survival: Wage Labor in the Late Twentieth Century* (coeditor)

*Oil: A Time Machine*

*Alternative Theories of Competition: Challenges to the Orthodoxy* (coeditor)

# A Prelude to the Foundation of Political Economy

## Oil, War, and Global Polity

*Cyrus Bina*

palgrave
macmillan

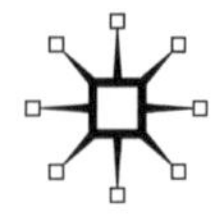

A PRELUDE TO THE FOUNDATION OF POLITICAL ECONOMY



First published in 2013 by
PALGRAVE MACMILLAN®
in the United States—a division of St. Martin's Press LLC,
175 Fifth Avenue, New York, NY 10010.

Where this book is distributed in the UK, Europe and the rest of the world, this is by Palgrave Macmillan, a division of Macmillan Publishers Limited, registered in England, company number 785998, of Houndmills, Basingstoke, Hampshire RG21 6XS.

Palgrave Macmillan is the global academic imprint of the above companies and has companies and representatives throughout the world.

Palgrave® and Macmillan® are registered trademarks in the United States, the United Kingdom, Europe and other countries.

ISBN: 978–0–230–11561–3

Library of Congress Cataloging-in-Publication Data

Bina, Cyrus, 1946–
A prelude to the foundation of political economy : oil, war, and global polity / Cyrus Bina.
pages cm.—(The economics of the Middle East)
Includes bibliographical references.
ISBN 978–0–230–11561–3 (alk. paper)
1. Petroleum industry and trade. 2. Power resources. 3. Globalization. I. Title.

HD9560.5.B477 2013
327.1—dc23 2012032613

A catalogue record of the book is available from the British Library.

Design by Newgen Imaging Systems (P) Ltd., Chennai, India.

First edition: February 2013

10 9 8 7 6 5 4 3 2 1

FOR ALL KNOWING SOUSAN
—Affectionately

*I have seen tempests, when the scolding winds*
*Have rived the knotty oak, and I have seen*
*The ambitious ocean swell and rage and foam,*
*To be exalted with the threatening clouds;*
*But never till to-night, never till now.*
*Did I go through a tempest dropping fire.*
*Either there is a civil strife in heaven,*
*Or else the world too saucy with the gods*
*Incenses them to send destruction.*

—William Shakespeare
*Julius Caesar* (Act I, Scene 3)

# Contents

# Note on Sources

With respect to published materials, although the author (Cyrus Bina) is copyright holder of the following work, with expressed permission of the publishers, the inventory of sources used (either wholly or in part) in this volume is as follows:

"Internationalization of the Oil Industry: Simple Oil Shocks or Structural Crisis?" *Review: Journal of Fernand Braudel Center* 11 (3), 1988, by *The Fernand Braudel Center* at Binghamton University, Binghamton, New York; "Some Controversies in the Development of Rent Theory: The Nature of Oil Rent," *Capital & Class* 39, 1989, by Sage Publishers, Thousand Oaks, California; "Limits to OPEC Pricing: OPEC Profits and the Nature of Global Oil Accumulation," *OPEC Review* 14 (1), 1990, by Wiley, UK; "Oil, Japan, and Globalization," *Challenge: A Magazine of Economic Affairs, Challenge: The Magazine of Economic Affairs* 37 (3), 1994 and "The Globalization of Oil: A Prelude to a Critical Political Economy," *International Journal of Political Economy* 35 (2), 2006, by M. E. Sharpe, Armonk, New York; "OPEC in the Epoch of Globalization: An Event Study of Global Oil Prices," *Global Economy Journal* 7 (1), 2007, by Berkeley Electronic Press, California; "Competition, Control and Price Formation in the International Energy Industry," *Energy Economics* 11 (3), 1989, by Elsevier, New York; "War over Access to Cheap Oil or the Reassertion of U.S. Global Hegemony," *Mobilizing Democracy: Changing the U.S. Role in the Middle East* (1991), Managing Editor of Common Courage Press, Monroe, Maine; "The Rhetoric of Oil and the Dilemma of War and American Hegemony," *Arab Studies Quarterly* 15 (3), 1993 and "On Sand Castles and Sand-Castle Conjectures: A Rejoinder," *Arab Studies Quarterly,* 17 (1 and 2), 1995, by the editor; "Towards a New World Order," *Islam, Muslims, and the Modern State* (1994) by Palgrave Macmillan, UK; "Is It the Oil, Stupid?" *URPE Newsletter* 35 (3), 2004, by Union for Radical Political Economics; "The American Tragedy: The Quagmire of War, Rhetoric of Oil and the Conundrum of Hegemony," *Journal of Iranian Research and Analysis* 20 (2), 2004 and "Crossroads of History and a Critique of

Prevailing Political Perspectives," *Journal of Iranian Research and Analysis* 26 (2), by the editor; *Oil: A Time Machine*, Linus Books, New York, 2012; and "Synthetic Competition, Global Oil, and the Cult of Monopoly," in *Alternative Theories of Competition*, Routledge, Abingdon, New York, 2013.

# Acknowledgments

This book carries a belated pledge made to Robert Brenner a few years ago at the UCLA faculty club. In the early and again in late 2000s, I was fortunate to spend my sabbaticals at UCLA Center for Social Theory and Comparative History with Bob and Tom (Mertes). Early in that decade, the US invasion of Iraq was imminent, as was its outcome that was then crystal to me. I asked Bob about the priority of writing a book on either oil or globalization. He looked me in the eye and said, you write a book about oil. Here, I wish first and foremost to express my gratitude to Robert Brenner who has instigated this volume.

This book was written in the course of one of the busiest periods in my academic life. In addition to a challenging teaching load and serving on several demanding academic committees, in the second half of the 2011–2012 academic year, from January to May alone, I had participated in several timely sessions, including ones arranged by Economists for Peace and Security (EPS), and chaired a session organized by URPE on the "Arab Spring," at the ASSA meetings in Chicago; a special conference on Constitution in Iran at the UC-Santa Barbara, which had given me also a chance to meet many distinguished scholars of Iranian studies, including Janet Afari, Hamid Akbari, Richard N. Frye, and Fakhreddin Azimi; a weeklong of lectures on globalization of oil (and theoretical basis of nationalization) in Buenos Aires, Argentina, at the invitation of the Universidad Nacional de Quilmes and Government of Argentina (in the company of eight other leading economists); and finally accepting an invitation to give a public lecture (accompanied by several class visits) on oil, US foreign policy, and globalization at Bentley University, Massachusetts.

I hereby wish to extend my appreciation to Andres Stamadianos Guzman for being such a great host and spectacular organizer during my stay in Argentina; I must also report that Andres is no stranger to the complexity of rent theory and its application to oil.

I am also grateful to Nader Asgari, Bryan Snyder, and Michael A. Quinn of Bentley University for their collegiality and hospitality during my visit in Boston.

I wish to express my gratitude to Chair of the Board of Directors of Economists for Peace and Security James K. Galbraith for extending the honor of being a Fellow of EPS, thus serving on the governing body of the organization. This of course formalizes my nearly four decades of laboring for the cause of peace and justice in the world.

I cannot conclude this acknowledgment without reiterating, again and again, my intellectual debt to three eminent scholars on both sides of the Atlantic, namely, Anwar Shaikh of the New School for Social Research, New York, and Ben Fine and John Weeks of the School of Oriental and African Studies of the University of London.

Last but not least, I wish to thank my longtime friend and colleague Charles Davis of Indiana University for offering to read the manuscript in its entirety and giving valuable suggestions that improved the book; I also want to thank Chuck for his sweet spot for building a judicious society free of exploitation.

Finally, I am delighted to acknowledge the support I received from all those at Palgrave Macmillan, among whom Leila Campoli's stands out; she did a brilliant job of getting this book beautifully through the publication. The usual disclaimers apply, however, as I alone am accountable for any errors of omission or commission in this book.

# Introduction

*If you would have a thing shrink, You must first stretch it…*

—Lao Tzu
*Tao Te Ching*

## I

At the intersection of two important questions of our time, namely global peace and sustainability of the planet Earth, stands a commodity beyond all commodities that has wreaked havoc with the material and mind of humanity in recent times. This commodity is none other than oil, whose *realities as well as its fictional impression* have both made a consequential impact upon public opinion today. In this book, we hope to demonstrate what is right and what is wrong with learned opinion across the board, before showing what is amiss with popular perception.

As shall be demonstrated in this book, oil is not an object but a trajectory, indeed a constellation of exigencies, events, actions and reactions, disputes and refutations, disparity and deviation, and, above all, contradiction and conflict across historical time and social relations fused and conjoined. This can be seen from the early development of oil in full-blown cartelization within the International Petroleum Cartel (1928–72) through to a competitive globalization beyond borders of any one nation-state by the beginning of third quarter of last century. This is the story of old colonialism bleached in neocolonialism, with all intents and purposes, and carried over and conveyed by the schizophrenic rubric of Pax Americana, before becoming history in the past tense. This was the end of a history and the beginning of a new one. In this context, oil crisis of the early 1970s, which has transformed petroleum (and energy as a whole), should have a pertinent historical place on its own right by virtue of its image in the global polity and economy of today.

The 1973–74 oil crisis was not an ordinary (or, shall we say, periodic interruption and renewal) disruption; this was not only a mother of all crises within its own specific socioeconomic configuration, but a part of a larger crisis in a series of turmoil and instability that inaugurated and ushered in the beginning of the end for Pax Americana (1945–79). What is also important in this context was the symbiotic relationship between the great powers and worldwide cartelization of oil. In other words, this is a concrete and tangible ingredient of a sound theory of imperialism, at the intersection of state and social relations. This has been demonstrated rather concretely and amply in this book through oil. But what is also demonstrated with similar concreteness and intensity is the decartelization and globalization of oil that took center stage in the early 1970s, and led to a full-fledged global crisis beyond oil. Indeed, in particular oil was part and parcel of a series of economic and political crises that had already begun in the mid- to late 1960s, which by then had unfolded beneath the superstructure of order that thrived, with some degree of assurance, for more than three decades in the postwar era.

This was the beginning of the unraveling of the weakest link in the chain that once was a pillar of stability, namely, the client-state segment of the Pax Americana. But this was also symptomatic of rifts within the more sovereign segment of Pax Americana. This is why the presidency of Jimmy Carter should be seen as the last hurrah for America. And that is how, despite the Reagan administration's ideological parade, Hollywood-style propaganda, and its dutiful pretense, the United States walked willy-nilly into an insubordinate and uncharted zone unlike that of its habitual own, and tended to wreak havoc across the divide of the then global polity. Intoxicated by the overpowering euphoria caused by collapse of the Soviet bloc, the post-Reagan America also had little chance for soul-searching to perceive the specter of time and to read what is written or unwritten on the proverbial wall. The euphoria provoked by the collapse of Soviet bloc in the early 1990s dulled public perception that America's formidable industrial base had already been unraveled piece by piece, plant by plant, and industry by industry just within a decade. This was surely the end of a history but not in the intended fashion of Fukuyama's jovial ruse (Fukuyama 1992). The irony of history in our scenario, in Fukuyama's witty rendition, would make the "last man" the very last US president in charge of the now defunct Pax Americana.[1]

In a nutshell, under the veneer of "optimism" attributed to Ronald Reagan, there was also the age of post-Bretton Woods, decartelization and cutoff of the umbilical cord of US foreign policy of oil, wholesale plant closings, massive privatization and outsourcing, and, not to mention, a remarkable class polarization in the United States. In a larger context, awful events of 9/11 were a visible and powerful shot across the bow, not so much for the attack by a crafty bunch of barefooted, discrete, non-English-speaking natives from a faraway land, but for the fact that the arbiter of time let them loose. Yet, perpetrators of 9/11, aside from foreign policy considerations, have accomplished a more sinister mission, which advertently or inadvertently struck at the heart of civil society in America. In Gramsci's terms, they pitted "political society" against "civil society" in America, by prompting the former to outdo the latter by the agony of constant surveillance across the homeland. They tossed the Trojan horse of blanket suspicion right in the middle of America's civil society (and, by implication, in the middle of other societies that had once been members of the old imperial club) in an atmosphere of fear and intrigue unleashed and utilized by the state. Hence the birth of a full-blown *paranoiac state*, which is by far more dangerous and self-destructive than the witch-hunts of the McCarthyist era ever were (see Aaronson 2011).

The following passage from the front page of the *New York Times* is merely a glimpse of what has become of posthegemonic America:

> The question is whether the Pentagon and military should undertake an official program that uses disinformation to shape perceptions abroad. But in a modern world wired by satellite television and the Internet, any misleading information and falsehoods could easily be repeated by American news outlets. The military has faced these tough issues before. Nearly three years ago, Defense Secretary Donald H. Rumsfeld, under intense criticism, closed the Pentagon's Office of Strategic Influence, a short-lived operation to provide news items, possibly including false ones, to foreign journalists in an effort to influence overseas opinion. Now, critics say, the missions of that discredited office are quietly being resurrected elsewhere in the military and in the Pentagon. (Shanker and Schmitt 2004: A1, A12)[2]

If this is not an indication of desperate reaction against the loss of hegemony (in an organic sense of the term), then we truly have no clue whatsoever as to what is taking place in this interim polity today (also under the Obama administration) before our eyes.

## II

The evolution of oil embodies the major structural changes that propelled the world from its late nineteenth-century socioeconomic posture toward the early twentieth century. Oil has not merely been a fuel of choice but a material necessity that preconditioned the concrete trajectory of capitalism in its unfolding across national and transnational boundaries toward its worldwide sweep of today. Therefore, the dialectic of oil and modern capitalism is more or less sufficient for the identification of capitalism and capitalist social relations. Oil is also a complex subject that unites the geography of production and accumulation capital in a historical synthesis. That is why we need to trace the development of the oil sector along the specific evolutionary stages in which the context of continuity and change interweaves with the dynamics of world development at large. In other words, not only did twentieth-century capitalism bestow upon oil a critical context that stuck with it until today, as a source of energy oil also bequeathed the latter with further identification—thus the phrase: "hydrocarbon capitalism."

One of the characteristics of oil in its early exploration and production has been the requirement of large capital investments for exploratory activity associated with unexplored fields surrounding new oil reserves, and costly development expenditures that are subsequently needed for extension and expanding of such fields once they were explored. Therefore, the evolution of the oil industry had not been and cannot be treated in a manner of a mom-and-pop enterprise in which capital has yet to turn into a well-developed process of concentration and centralization. On the other hand, in the late nineteenth century, Taylorism was just giving rise to standardization and thus automated assembly line mass production in need of capital on a scale beyond individual wealth. That is why oil was characterized by the assemblage of several financial syndicates for the venture of exploration in both the United States and abroad. And it is the minimum size of capital that in part plays a pivotal role in development of capitalist competition in oil and in other businesses. The genesis of hydrocarbon can be traced to colonial fusion of capitalistically developed and undeveloped parts of the world—a world whose overwhelming majority had not yet lived within capitalism proper. The evolution of oil has taken a course in a century-long development (1870–1970) that eventually came a full circle to embrace the entire globe (and global capitalism) and to lead the way to what is known as the era of globalization. Thus, for oil the twentieth century is the age of growth and maturity, from a limping capitalism—on cartel's crutch—to a capitalism that walks upright, as social relation, on its own across the planet.

This starts with from the 1846 Bibi-Haybat discovery of oil in Baku, Caucuses, and the subsequent 1859 Drake's strike in Pennsylvania, and oil's international evolution beyond the United States in the West, and Caucuses in the East, obtaining its focal point within the region that due to British colonialism has since been known as the Middle East. The evolutionary development of oil and its prominent shift to this region is thus a significant part of the story that belongs to the first stage in the evolution and cartelization of oil worldwide. In other words, D'Arcy's oil concession of 1901 in Persia led to discovery of oil in Masjid Suleiman (southwestern Persia) in 1908, and the subsequent colonial (and semicolonial) economic and political domination of Britain. This, of course, was a cornerstone upon which the emerging powers, such as the United States, gained foothold on the precious yet shifting sands of the Middle East until the globalization of oil in the early 1970s. The evolution of oil has gone through a second stage that we identify as the transitional period of 1950–72, a transition that eventually leads to a full-fledged decartelization and globalization of oil via the oil crisis of 1973–74.

The role of the 1973–74 oil crisis is pivotal in revealing the tip of decartelization and subsequent globalization of oil. This includes the illustration of an evolutionary mechanism that includes competitive worldwide pricing of oil against unequal costs (and productivities) of the various oil-producing regions. As is demonstrated, the crisis indeed has led to decartelization of oil in all oil regions of the world—including US oil and at the same time, through Organization of the Petroleum Exporting Countries (OPEC) (as a competitive context in spot and future markets for all oil irrespective of geographical location). The kernel of confusion on the question of oil is not the difference in conceptualization between the Right and the Left. It is rather the similitude of imagination that makes them party to the spread of misinformation and cover-up on the alleged causal relation of oil and war in the present era. It shall be demonstrated that, by imitating the right-wing notion of competition and its idealist (i.e., axiomatic) spectrum of pure competition/pure monopoly, the liberal/radical Left succumbs to a nostalgic theory that still describes the present according to the past history under the aegis of cartelized oil.

## III

We speak of crisis as the process of renewal and restructuring. We wish to clarify that renewal and restructuring as an entity applies to the continuity and discontinuity within the same system and along similar trajectories. For instance, the periodic economic crises in capitalism fit this description fully. But when we speak of the 1973–74 oil crisis, we should be cognizant of the fact that there was neither an intent nor a possibility for renewal or restructuring of the same cartelized, neocolonial system, while transnationalization of social capital had disrupted and annulled the arbitrarily administered pricing scheme that was in place within one of the most notorious cartels in the history of capitalism. Therefore, the 1970s oil crisis was not about renewal but burial of the old. As is demonstrated, particularly in chapter 4, the road to competitive globalization of petroleum had to run through nationalization of the oil deposits in the Middle East, Africa, and Latin America in the mid-1970s. These nationalizations, intended or unintended, had nothing to do with recartelization of oil but instead were driven by valorization of landed property (i.e., the ownership of subsoil oil deposits) under capitalism, and thus should be deemed as both commensurate and compatible with competitive globalization of oil.[3]

The April 16, 2012 renationalization of oil in Argentina, which also had our full support while in Buenos Aires a month earlier, follows the same rule. The burial of the old system was not an exclusive objective of Libya, Algeria, Iraq, or Venezuela at the time. It was also the aim of so-called US oil independent companies as well.[4] There was also a question of the decline of US domestic oil fields that had long been overutilized by consecutive and unwarranted investment of capital at the *intensive margin*, by the International Petroleum Cartel (IPC), which were in desperate need of rationalization. This shows the connectivity and interdependence of a modern industry that was coming into its own by shedding its cartelized shell and departing beyond its colonial and neocolonial legacy. This was a metamorphosis of the first order. The development of OPEC, from its inception, in 1960, through its infancy and eventual maturity, is also anticipated within the contours of this transformation. OPEC cannot maneuver as it wishes without the validation of globalized oil, which puts limits on any discretionary action or demand of the former (see Bina 1985, 1990, 2006; Bina and Vo 2007, and chapter 3 in this book).[5] The evolutionary history of oil thus provides a useful glance at the epochal transformation of this sector, from cartelization to globalization, which in turn upset the applecart and overturned the traditional view of oil and *geopolitics*. Globalization of oil is merely a feature of mature capitalism with respect to valorization of oil deposits across the globe.

As articulated in chapter 1 the oil crisis is not a consequence of conspiracy; neither was it explicable by "dependency theory" via "OPEC offensive" and/or purported "Third World" nationalism, nor intelligible to mainstream economics orthodoxy. The 1973–74 oil crisis has remained an enigma to many orthodox as well as heterodox economists, and international relations specialists, to this very day. Therefore, it is a small wonder both the Nixon and Ford administrations had not been able to decode and grasp, even for their own sake, the shift of tectonic plates of decartelization of oil, and callously blamed it on their own dedicated lieutenants and gendarmes (i.e., the custodians of imperialism) at the helm of assorted client-states in the Middle East. The most pestered of these was the Shah of Iran—one of the two "pillars" of Nixon Doctrine in the Persian Gulf and the finest and most obedient son of Pax Americana—who had to take it on the chin from Gerald Ford, one of the most oblivious US presidents in the history of the republic. Ford did not fathom what hit the system; this was alike for Henry Kissinger who thought he knew. The Carter administration had no clue either as to what happened to oil. Carter's foreign policy team opted for the "Rapid Deployment Force" against their loyal and struggling allies. If this is not self-destruction, we do not know what self-mutilation means.

## IV

Social relations underpin the overall dynamics of capital accumulation from which the indispensable tendencies of social capital emanate. The content of these relations, however, are historically specific to capitalism. The two essential hierarchical levels within social relations in capitalism are (1) the overall social structure and (2) the congruent institutions at each social stage. Structure and institutions both encompass the domains of economy, polity, and civil society. In a simple analogy, the distinguishing feature of socioeconomic structure in capitalism, in comparison with other historical systems, are perhaps parallel with the basic structure of DNA in human species as opposed to that of other species. The institutions, on the other hand, may be considered as the evolutionary forms and variations of the basic theme. Yet, to capture the dynamics of mutual interactions of basic structure and institutions in capitalism, it is necessary to conceptualize and examine an *evolutionary stage theory* of capitalist development. As a result, social relations are the manifestation of definite, historically determined structural causation and thus structural transformation in capitalism (see Bina and Davis 2008).[6]

The conquest of mode of production is not so much about establishment of the world market and "the chase across the globe," which to some extent was taking place during Marx's own time; it is rather about blanket valorization of capital and operation of the law of value in every nook and cranny of the planet. Therefore, speaking of globalization in the late nineteenth century is not only deceptive from the standpoint of historical materialism but also watering down the meaning of today's globalization. Globalization is not a policy but a stage in the development of capitalism in which there is nearly no place to hide from the omnipresent grip of capital—as a social relation. The epoch of globalization of social relations is equivalent to completion of conquest of the mode of production anticipated by Karl Marx. This epoch is also about encroaching on environment (and ecosystem) that is hanging in the balance. Globalization is about instantaneity and spontaneity of forces that are beyond the regional, national, geographical, and legal boundaries. And there has not been any time in the recorded history of capitalism that capital had been so seamlessly stretched out in hegemonic fortitude.

Imperialism according to olden connotations, including V. I. Lenin's (as the highest stage of capitalism), idealizes capitalism proper based upon the undeveloped phase in which four-fifths of humanity had neither lived under capitalism nor even perceived as to what it does. This is not short of doing away with historical materialism. Since the early 1920s, neither has capitalism stood still in its embryonic form nor has competition rested in its undeveloped, frozen mode prevailing at the turn of the previous century. We have passed the milestone of complete socialization of production across the globe, and there is no turning back. The threshold of transnational valorization of capital, that is, globalization of social relations, has been met. As an epoch, the substance of Leninist conditions for imperialism has already been overtaken by the omnipresence of Marx's value theory concretized in formidable and robust dynamics that now so distinctly negate the purpose and conduct of imperialism; the negation of an outmoded epoch by the conquest of mode of production as globalization—all within capitalism. The postwar system of Pax Americana, therefore, can be viewed as a transitory period that connects the era of Lenin's Imperialism to the epoch of globalization. And if the Pax Americana was an epoch of transition, then, we are now in transition from this transition.

Given the above framework, despite the fact that capitalism had already turned into a social relation metastasized and manifested in every nook and cranny of the globe, there are activists and scholars (across the political spectrum) who rather insistently view the evolutionary story of oil through their rear windows. They erroneously think that by grabbing oil—as in colonial times—by military force one may reverse the course of history. There is another side to this double blunder. Against the omnipresence and awesome connectedness of today's capitalism, there are governments, groups, and individuals that think "energy independence" is possible with some diligence and perhaps through a bit of "American ingenuity." In this contest, the liberals and leftists insist on conservation and/or renewable energy, which by themselves are worthy projects. The right-wingers, on the other hand, wish to go after every drop of oil regardless of consequences—the demise and degradation of the environment and the dilapidation of quality of life for their own grandchildren. The Left is living with fantasy; the Right is living for profit.

## V

We also hold that the crisis of hegemony does not end with the renewal and restructuring of the now defunct Pax Americana, as the global polity has long-crossed the Rubicon riding on negation of American hegemony with no turning back. As we articulate in chapter 7, "Hegemony is a mutual characteristic of the system as a whole—not a separate property of the hegemon . . . Hegemony thus thrives through reflection of the *whole*, not exertion of the parts." In the absence of Pax Americana, speaking of US hegemony is in the past tense. One of the significant differences between today's economy and polity, and the one under Pax Americana is the demand for spontaneity—the spontaneous diffusion of global political power, despite US infatuation with voluntarism, exceptionalism, unilateralism, and preemptive politics in the global affairs. Yet, no amount of unilateral intervention or selective politics in pursuit of power would reverse the hand of time—"what is done cannot be undone."

And if the history of the past few decades is of any consolation, given the tsunami of epochal change, any attempt at turning back the clock would not be without repercussions that may boomerang and accelerate the fall further and thus wear down the capacity of new global polity in the making. This is what one may roughly label as switching the agenda on the future polity by the United States—by default and by self-induced setback. It should be simple that once the train of new epoch had departed from the old station, attempts by nostalgic passengers, who try to run backward through the rear cars, may not be received kindly by forward-looking passengers of the new era—the global majority. This may disturb the peace, for a while, but it could also win over the many injured and undecided passengers who now wish to get their hands on these self-defeating time-travelers from the past. Hence time is not on the side of the United States, neither are preemptive actions carried out in the name of global leadership. Hence the power that is in decline is more dangerous to peace and stability than the one that is on the brink of ascendancy. Commencement of the third millennium has not been kind to the United States, despite its self-image as the only superpower in the world. This self-image was in part a corollary of the precipitous collapse of the Soviet bloc, which then provided the United States with political victory and opportunity for undeterred unilateralism.

To say that, after the collapse of Soviet bloc, the United States has become emboldened by utilizing the "vacuum" that was generated by the absence of a formidable adversary in the globe, is not insignificant on the face of it and from a *mechanical* point of view. There is no doubt as to the jubilation and acceleration of US belligerence since the aftermath of Soviet fall. But from a dialectical view—that focuses on transformation from within—the *vacuum analogy* does not go very far. In other words, if the change comes from within the apparent state of vacuum might not remain so, as the change in quantity turns to change in quality. To fill (i.e., occupy) the apparent vacuum (by an entity alien to change) thus betrays the original cause of the transformation in the short run; yet in the long run the newly developed quality may eventually liberate its domain. The analogy of power vacuum attributed to the US behavior is an impeccable catalyst but not a kind that could be utilized for the explanation of the original cause of epochal change. This may explicate the universality of change that had also led to implosion of the Soviet bloc.

The change in nature and its social counterpart always comes from within. The slate is not clean even in its strict primordial sense nor the stage empty of social, political, and ideological embryos. The change also develops through contradiction—not harmony. But not any and every change is the change of an order to the next. We need to look at synthetic nature of social relations and politics. But the change itself is a living creature that incubates in any and every standing structure. Yet, water does not on its own climb. And that is why there is class struggle beneath any and every turned and unturned stone in today's polity—from Tunisia, Egypt, Yemen, Bahrain, and Syria to the epicenter of capitalism on Wall Street. The color and flavor of "nationalism" in many of these struggles might be enchanting to the many observers or even to the many who found courage and civility to participate in them. But flavor and color do not help pay one's bills. Yet, one has to grapple also with the paramount political forces that knocked down the very last vestiges of Pax Americana by defacing native faces of the now defunct American hegemony. Finally, paying lip service to the cause of the Arab revolt and robbing it of authenticity, while giving the Obama administration a pause or two to catch up with post-Pax Americana's house of cards in the Middle East and North Africa, neither will the face of change disappear nor is the hypocritical US foreign policy lost on the inhabitants in the region. Thus the effect of all this gives us a window of opportunity to appreciate where the post-Pax Americana world is heading.

## VI

International relations specialists—particularly those on security—have been quick to seize the security of oil even after the stage of globalization and objectification (i.e., reification) since the 1970s. These specialists act not unlike cops-on-the-beat with little understanding that, in the world of seamless interdependence of oil supply, one has no need to mimic a kind of control once exercised by the IPC in the earlier neocolonial era. These specialists show little awareness about the epochal transformation of global polity, let alone the epochal transformation of oil. They are not as quick as, say, tellers at the bank counter who are now cognizant of the fact that their position en masse has already been transferred to a machine that automatically disburses money 24/7 with no appreciable security problem.

Within orthodoxy in the field of international relations, the scholars who often rely on the "OPEC offensive" image of the 1970s, miss the boat on the grander transformation that combined OPEC, the US domestic oil, and all other oil (and energy) producing regions under one indivisible rule. Being trained in fictional constructs symbolized as "nation-state" and "anarchy," dubbed "realist" approach to the international politics (Krazner 1978, Drezner 2011)—or tutored to opt for control in an old colonial style with a new garb as "neoconservatives"—many of these political scientists fail to recognize that OPEC is neither a cartel or a monopoly; they fail to distinguish rent from profit in order to decipher the puzzle of concentration (and centralization) of capital accompanied by the overpowering presence of hypercompetitive and globalized oil, notwithstanding OPEC oil rents.[7] Their timeless concepts are punctuated by improvised references to assorted transhistorical events—from the Roman Empire to romanticized American history—in panoramic fashion. Finally, on the issue of US global leadership, some of these specialists are so prejudiced as to not let go of the past even for the sake of global peace and stability.

This anachronistic orientation also plays a part in heterodox international relations literature in which the circular notion of power unites with the transhistorical notion of hegemony. The only recourse for grounding the theory is either to resort to sheer empiricism (and ad hoc descriptive history) or to fall back on mainstream economics as to where and when the question of oil should take center stage. This predicament remains with specialists who cautiously, for depth and consistency, capitalized on social theory but have to grapple with the issue concerning capitalist competition and the globalization of oil; they hang on to monopoly theory (and cartel) and thus legitimize US foreign policy of yesteryears as if the IPC were still alive and kicking. These liberal/radical scholars also have not yet awakened to the reality of today's global polity. As a consequence, a wrong turn on the foundation of political economy can be costly on both theoretical and policy grounds. Bromley (1991, 2005) provides a quintessential example of this category. We read:

> The United States does seek to exercise a degree of influence over world oil second to none, but the *form* of that influence is very *ambiguous* and very different from the kinds of control over raw materials traditionally associated with imperial powers (Bromley 2005, abstract, emphasis added).[8]

A wrong turn on the notion of competition, combined with eclectic methodology with respect to the theory of value, provides false hope when falling back on monopoly oil in conjunction with American hegemony. Hegemony (aided by misconception of "monopoly oil") conceives and strengthens the question of oil security, not so much for random contingencies but rather because of the combination of the faulty vision of American leadership and anachronistic interpretation of oil. And it is ironic that such categorical (nuanced or otherwise) verdicts should come at the time that the umbilical cord of US foreign policy is cut off, as the Trojan horse of IPC is already six feet under and resting in peace, and the fractured client-state subsystem under the now defunct Pax Americana is now a museum piece. It does not occur to some of these writers, Bromley included, to return to the list and composition of oil contracts, given to transnational oil companies in "post-US withdrawal" Iraq, and to see whether the so-called American companies are exclusive winners. But this writer is not so naive as to think that this would settle the argument with these entrenched antagonists. They always turn around rather preposterously and say, "because the United States, the alleged hegemon, is watching over the interests of everyone in the game." And one has to go back to the drawing board again and again to demonstrate that king is already dead—long live the king!

A vivid example of the loss of American hegemony can be seen in the recent vote in the UN General Assembly on the non-Member State status of Palestine and the absolute abandonment of the United States by virtually all its allies who either stepped aside or voted against its predictable, prejudiced, and, indeed, pitiful stance in support of Israel (see UN General Assembly 2012). What the Obama administration did prior to the vote was to gather a teeny contingent of invented states (with the exception of Canada), including Marshall Islands and Micronesia. To employ Donald Rumsfeld's phraseology, only a "dead-ender" would do such an injury to its footing, particularly in the time of its vanished hegemony.

As will be elaborated in this book, the transformation of oil—from cartelization through to decartelization—has been parallel with ascendance and blossoming of American hegemony through to its eventual downfall. Each of these transformations is linked systematically to a power structure that is embedded in valorization of landed property (i.e., of subsoil oil deposits), including the role of states that hold those deposits. The result is an embedded power relation that has been mediated through the globalization of oil. This speaks to the redundancy of US hegemony, particularly where it comes to "security" of supply (of oil). As is demonstrated throughout this book, the "law of value" acts as the biggest *Robocop* on the block along with its own omnipresent globalized oil market—an objective (and evolutionary) replacement for subjective neocolonial outfits, such as the IPC. This, of course, shall not be free of short-run contingencies and periodic crises, which in either case is beyond the control of any mortal soul (or a cartel, for that matter) in the universe of capitalism. On the other hand, if speaking of "security of supply" is a bashful petition for exclusive security (read monopoly) for the United States—and a handful of members in the now defunct imperial club—then, this is surly a losing proposition that should embarrass those who advocate it (see Stokes and Raphael 2010).[9]

## VII

As demonstrated throughout this book, oil from the least productive oil region is entitled to a competitive profit. This reflects a normal rate of return on capital investments that, notwithstanding the risk and uncertainty, move rather competitively in and out of the industry on a regular basis. By comparison, oil production from more productive oil regions is entitled to a differential oil rent, in addition to normal profit. In this manner, the long-run price of oil is set by the US oil production price (US regional oil cost, plus competitive profit), which in turn represents the gravitational center of short-run fluctuations of oil prices worldwide. This also has considerable implications for the question of environment and the issues that are hanging in the balance in view of the popular but fictitious desire for "self-sufficiency" in oil and energy in the United States.

That is why, so long as the production from least productive US oil fields is to continue, the measly production from new explorations, such as from the US Outer Continental Shelf and/or Alaska's National Wildlife Refuge (i.e., more productive US oil provinces), would neither change the center of gravity (the long-run price) nor markedly reduce the short-run price of oil in the United States. Again, this is not because of the alleged "oil monopoly" invoked by popular wisdom, but because of the very fact that *differential oil rents* are an outcome of competition among the lesser- and more-productive oil fields; and least productive oil fields are merely entitled to competitive profit, without rent.

As for "peak oil," the connotation creates a kind of opacity, similar to a sleight of hand that conceals one's attention from the tautological reasoning underneath. This epiphenomenon relies heavily on the estimation of existing oil reserves that are directly dependent upon the state of technology, profitability of extraction, price of oil, long-run consumption pattern, and a myriad of other complex factors that critically relate to accumulation of capital globally. To all this one might add the question of universal uncertainty, its nonlinear dynamics, and the inapplicability of "probability theory' " with respect to its future trajectory.

Speaking of "energy independence," in an era of interdependence and unification of nearly all oil regions (including US oil), is little more than dreaming. In this context, strange bedfellows like right-wing conservatives and left-wing liberals and radicals have now joined forces for "clean energy" and "energy independence" from "monopoly" of the Middle East oil. Being the highest cost region of the world, US oil has regulated the production price of world oil since the 1970s, thus constituting an indivisible part of the global petroleum industry. An organic unity imposed by globalization precludes national solutions via isolationism, romantic localism, fear-mongering Malthusianism, peak-oil charlatanism, and xenophobia. As we demonstrate in this book, production of oil from the least productive oil region is entitled to a competitive profit. This reflects a normal rate of return on capital investments that are competitively bound to move in and out of all industries on a regular basis. Hence, opting for "self-sufficiency," say, through further exploration in the US Outer Continental Shelf and/or Alaska's Arctic National Wildlife Refuge (ANWR), would neither reduce overall production costs of US oil nor add enough supply to change the magnitude of world production price to be worthwhile.

Given that the anticipated production price of oil (per barrel) from ANWR would be considerably below that of US domestic oil fields in the lower 48 states, these newer fields will generate differential oil rents plus profit, whereas existing oil fields would retain their status as before with respect to regulating long-run price of oil. This gives considerable exposure to the purported option of "self-sufficiency" (particularly, in its happy-go-lucky "drill, baby, drill" phraseology, see Harkinson 2012), which in turn leads to tremendous environmental impact and multiple chains of events that may not submit to any calculable risk whatsoever. Thus, silly demand for "self-sufficiency" in oil and energy, together with sinister reasoning for further domestic drilling for oil in the wild, would neither produce a significant supply of oil nor reduce the long-run price of oil in the United States. Shifting from nonrenewable to renewable energy is a fine idea that should not be mixed up with the hoax of "energy independence." Therefore, given the excessive use and abuse of energy sources, which causes environmental degradation, the indivisibility of life on the planet requires universal remedies beyond pitiful nationalism and utopian self-sufficiency.

On the other hand, a glance at recent US natural gas production by "fracking"—otherwise known as hydraulic fracturing—may add costs (private and social) that are high, hidden, and exceedingly uncertain to be sustainable in the long run (see Marks 2012 on the hazard of fracking). Besides, globalization of oil is a concrete example of an interdependent world that progressively advances toward further fusion and mutuality. Therefore, penchant for "energy independence" is not something that is dependable or maintainable in today's globalized world. The appeal to such idealized alternatives, though a populist one, is inundated with xenophobic overtones and an inexplicable sense of nostalgia. The sore point is that the alleged reliance on Middle East oil and OPEC "monopoly" always find a dramatic rationalization in the fable of "perfect competition." As can be seen, our argument is not limited to esoteric academic issues; the framework presented here very much concerns two very critical arenas of our public life, namely, US environmental policy (and climate change) and US foreign policy.

# 1

# World Oil and the Crisis of Globalization

*Truth emerges more readily from error than from confusion.*

—Francis Bacon

## Introduction

The oil crisis of 1973–74 was the symptom of the underlying fundamental changes that forcefully led to the internationalization of the oil industry. The production and pricing of crude oil associated with the various oil-producing regions of the world have since become part of a unified process through global competition. It was a mother of all crises that led to the restructuring of all oil, thus brought cheap and not so cheap oil under one all-inclusive, globalized market. This prompted the collapse of the International Petroleum Cartel (1928–72); this included the intricate basing-point pricing system at the Gulf of Mexico and the Persian Gulf, and what lingered as the institutional wherewithal and purposeful paraphernalia linked to the cartel's success. In the meantime, the "Postwar Petroleum Order"—an indispensable part of the international order of the Pax Americana (1945–79)—had begun to fall by the wayside, and the umbilical cord of the US foreign policy was cut from cartelized oil for good. The crisis appeared as a faint signal at first. This prompted the United States and its habitual Western alliance, and the titans of the International Petroleum Cartel (IPC), to engage in an old mode of diplomacy and negotiation to find a customary solution. But it soon became clear that not only the Middle East oil but all oil across the world had also crossed the Rubicon—the new era was about to begin. A cauldron that had been bubbling for quite a while—perhaps since the days of nationalization of oil in Iran and the overthrow of Mohammad Mossadegh—blew its top like a gigantic volcano. The steely law of history seemed to have shown an ironic display of objectivity and resolve. The gusher of discontent was so vast, so dense, and so sudden that it took years to sink in even in the psyche of regulars who thought they had intimate knowledge of oil and politics. And, while bewildered or struck by a healthy bolt of amnesia, the vast majority to this date is still underestimating.

In contrast to the prevailing opinions at the time, the significance of the oil crisis was not due to the (temporary) shortage that resulted from the imposition of the embargo; rather, the oil embargo itself was the symptom that revealed an underlying transition that had already been taking place toward the globalization of the oil industry. One has to appreciate that the process of structural transformation in oil production had already begun in the late 1960s and early 1970s. The 1973 oil crisis was simply the culmination of that process, which ushered in an entirely new period in which an end was put to separate regional pricing, inadequate unification, and localized value formation within the global structure. The oil crisis of 1973–74 was not an ordinary oil shortage, similar to the ones that the world experienced in the 1956 Suez Crisis or the 1967 Arab-Israeli clash. This crisis was conveyed by a severe shortage resulting from the Arab oil embargo (see Akins 1973), but it was underpinned by socioeconomic/sociopolitical forces that had long been laid at work toward the persistent transformation of global order. Hence it would be naive to reduce the oil crisis of the early 1970s to its manifold effects and multifarious impacts such as the suddenness of supply interruption and shortage alone. In the wake of the crisis stood the restructuring of the entire oil sector from top to bottom, doing away with monopoly and allowing price determination through global competition—including competition between the least and the most productive oil regions of the world. These conditions, in turn, necessitated the formation of market prices that were based upon production costs of the least productive oil region, and the synchronized formation of differential oil rents in step with the existing productivity of oil fields across the various oil regions of the globe.

We shall demonstrate that the formation of differential oil rent came about through increased competitive conditions rather than through monopoly. We shall identify the US oil region as the least productive in the world, and show that during the period leading up to the crisis there was a significant decline in the productivity of the aging US oil fields. The increase in the cost of production of the least productive oil region together with the internationalization of oil production, led to the generalization of high market prices within the entire industry.

The first section is a critical review of the literature on the oil crisis. The second section examines the characteristic features of US oil production. The third section presents an alternative theory of the oil crisis. The chapter concludes with a summation and setting the crisis within a larger polity.

## The Review of Literature

The oil crisis of 1973–74 was an important economic and political event that has been embroiled in controversy ever since. At first glance, because of the diversity of opinions and the number of unsettled questions, there seem to be as many theories about the oil crisis as there are theorists in this field of inquiry. But it may be useful to distinguish the common threads among prevailing views on the subject to be able to discern various theoretical lines and schools of thought. We divide the prevailing views on the oil crisis of 1973–74 into three main categories: (1) traditional theories of the oil crisis; (2) dependency theories of the oil crisis; and (3) conspiracy theories of the oil crisis.

### Traditional View of the Oil Crisis

This category contains an extensive spectrum of arguments about the nature of the oil crisis of 1973–74. The analyses often contain references to such notions as the oligopolistic structures of the oil companies, the collective decision-making of Organization of the Petroleum Exporting Countries (OPEC), and the operation of supply and demand in the international oil market. Although many disagreements exist among these theorists, nevertheless, the majority tend to approach the oil crisis of 1973–74 more or less in the same theoretical and methodological fashion.[1] To sum up the arguments made by these theorists, one needs to link together a combination of emphases which add up to an explanation of the oil crisis of 1973–74. The first emphasis is upon "law of supply and demand," within the sphere of exchange (Vernon 1975a). The second emphasis is upon the element of monopoly and "ability" on the part of OPEC to set prices at will (Penrose 1975). Third, some speak of the dependence of the US economy on foreign oil, especially on OPEC oil, which in turn created the severe momentary shortage that is said to have threatened "the supply security," as the determining factor (Lenczowski 1975, McKie

1975, Blair 1976). Finally, it is often said that it was the suddenness of the price change and the problems of adjustment that resulted in the crisis; if there had been a possibility for a smooth transition, the oil crisis might have been avoided (Blair 1976). In addition, from the methodological standpoint, the crisis is either explicitly or implicitly considered to have resulted from the change in the perception of the actors involved in the oil market, rather than being the outcome of the changed realities of the time (Dasgupta and Heal 1979, Ch. 15).

From the viewpoint of supply conditions (Blair 1976), the shortfalls were short-lived after the oil embargo. In other words, a temporary shortage developed that, according to conventional economic theory, was similar to previous oil shortages and thus would not have any significant impact upon long-term equilibrium prices. As the history of this period vividly indicates, however, exactly the opposite resulted. But the conventional theory insists that the factors that established a higher floor for the post-embargo price of oil were the result of price determination by OPEC, and the dependence of the United States on imported oil. It should be noted from the outset that the first step in this analysis is to identify the cause or the causes of the oil crisis. Despite this necessity, most theorists in this category considered quite a few factors that were associated with the crisis without being able to understand the systematic relationship among them to identify the underlying causes (Bina 1985, Chs. 3 and 4). In the final analysis, for the majority of the theorists of this category, "OPEC-determined" oil prices, along with the notion of US "dependence" on imported oil, was considered to be the principal cause of the oil crisis.

There are, however, a number of misconceptions in this conclusion. First, the two mechanisms of the "posted price" and market price of oil are being confused. Second, price determination is considered to depend upon the will of OPEC, rather than understood as the outcome of objective production conditions. Finally, the conclusion tends to imply that the United States was unable to challenge OPEC because it had become a net importer and could no longer supply oil to the world market as much as it had during the 1956 and the 1967 Middle East conflicts that had also led to temporary interruptions. The question to be asked here is why, considering the condition of excess supply that prevailed following the removal of embargo, the oil prices did not decline significantly. If the laws of supply and demand did not hold in this particular case, then what kind of mechanism tends to regulate the process of price formation in this industry?

Another common feature of these theorists is that they have not developed a mechanism to connect the process of price determination in the preembargo period with that of postembargo conditions. The reason for this seems to be the lack of adequate theoretical and historical perspective. These theorists either relied on supply and demand conditions within the sphere of exchange, or resorted to the notions of monopoly, cartel, and oligopoly to describe the sudden price change of 1973–74 that affected the entire globe.

Not only after the oil embargo ended, but even prior to it, the market was also flooded with huge quantities of crude oil. Moreover, even the decline in demand resulting from the worldwide recession of 1974–75 did not seem to cause any substantial drop in the price of oil. In fact, after the oil embargo of 1973, a new floor for oil prices was clearly established. This implies that changes in supply and demand conditions were themselves the consequence of more fundamental changes in the international oil industry. In the final analysis, the traditional theorists argue, OPEC acted like a monopoly in setting the oil price unilaterally at its fourfold increase level. When asked why OPEC did not act in the same manner that it had in the previous years, most of these theorists reply that, in addition to the existence of monopoly, factors such as the US "dependence" on the foreign oil and the rising tides of resource conservation contributed to the crisis. The most explicit argument of this sort is developed by Dasgupta and Heal (1979) who state that all these developments resulted from changes in the perception of the actors involved and not the result of changes in the actual situation—a shallow argument.

There is a fundamental problem with the above formulation. It suggests implausibly that objective realities do not exist outside of one's subjective perception (Dasgupta and Heal 1979, Ch. 15). That is why these authors maintain that, in the absence of futures markets, it is difficult to know how the perception of buyers and sellers are formulated. What the "futures market" does, however, is to present the extent of fluctuations and not necessarily the indispensable changes that had occurred through the crisis since early in the 1970s—the fresh mechanism that led to founding of the value, differential rent, and prices beyond the IPC (1928–72) in the entire oil industry. This is only a glimpse of innocence and complete irrelevance of the mainstream economic theory to a critical question of our time.

In retrospect, one may observe a great deal of displacement in the interpretation of the traditional school in the determination of the underlying cause of the crisis that established a nearly fourfold price increase, and brought about a transfer of a significant amount of wealth in the form of differential oil rent to the more productive oil regions. But it should be realized that the majority of these theorists were initially correct in describing the events and circumstances that were unfolding during the crisis. For instance, the impression of the "OPEC monopoly," or the notion of "dependence" of the United States on imported oil, is a correct observation. But what is rarely acknowledged is the distinction between impressions (i.e., observable effects) and the primary causes.

As a result, these theorists were unable to explain the process of the oil crisis of 1973–74; at best they described the consequences of the crisis and its conditions. Their description of the supply-demand relation, the "posted-price" determination by OPEC, and the US dependence on imported oil is certainly true. What is also true is that these writers failed to pass beyond the surface of these facts to build a theory of the oil crisis.

## Dependency Theory of the Oil Crisis

The dependency theories of the oil crisis of 1973–74 are deeply rooted in dependency theory in general.[2] This chapter neither claims nor intends to examine all the issues involved with this theory (see Prebisch 1950, Frank 1969a, 1969b, 1972; Emmanuel 1972; Amin 1974; Girvan 1976; Weeks 1981b; see Brenner 1977, Weeks 1981b for a critique; for further investigation see *Latin American Perspectives* 1976, 1977, 1979, 1981). The task here is rather to flush out specific claims of the dependency methodology as it relates to the analysis of the oil crisis. Thus, to begin with, we shall focus on Girvan's (1975) evaluation of the oil crisis of 1973–74, together with evaluation of authors, such as Tanzer (1974) and Stork (1975), who had made contributions to the subject from the standpoint of the dependency theory. The major argument expressed by most of these writers presumes a hypothesized "OPEC offensive"[3] against the industrialized countries of the West to achieve self-determination and sovereignty. This "offensive," however, is said to be a reaction to the prolonged relations of domination that existed between the Imperialist countries on the one hand, and the Third World countries on the other. Such domination was accompanied by unequal relations, and consequently unequal exchange in trade, between the "center" and the "periphery"—a precursor of what is now known as the world system theory (see Wallerstein 1979). Some authors added that the atmosphere of the post-Vietnam era created a general political condition that permitted the dominated countries of the Third World, notably OPEC, to launch this "offensive" (Girvan 1975: 147).

The most fundamental error of these theorists is eclecticism, that is, putting the questions of politics, economics, geography, international relations, and epoch on an equal footing without seeking to determine their structural relationship and without any specificity as to the underlying cause(s) of the oil crisis. For instance, the process of price formation in the oil industry is identified with the determination of "posted prices" by OPEC. There is no consideration of the laws of motion of capitalism as an alternative system of value and price formation through competition in the event of collapse of the IPC. Instead the emphasis is upon determination by *monopoly power*. It is not surprising that the dependency view considers the oil crisis as an offensive against unequal exchange. Some even called it an artificial crisis, since at the end they did not perceive any change in the magnitude of "unequal exchange." And setting up the problem this way, the dependency theorists missed the boat on the globalization of oil and, more importantly, on globalization in general in the era of post-Pax Americana.

These writers have scarcely recognized that prices are the phenomenal form of values in production and that value formation emerges through competition. They have made a double error of contrasting monopoly to competition, and committing price determination to "market power" circularly and in the absence of a viable theory of value in the oil industry.[4] Discussing the notion of competition, these theorists tend to equate the existence of a large number of firms in an industry with competition, and that of a very few firms with monopoly—thus succumbing to economics orthodoxy. It should also be pointed out that such equivalence would be a clear misconception of both monopoly and competition in capitalism. Here one moves from competition to monopoly through moving along the spectrum of "pure competition"/"pure monopoly," a fanciful stratagem that has absolutely nothing to do with the capitalist system of production and exchange. These scholars have failed rather miserably to realize that competition and integration in capitalism are part and parcel of a synthesis, thus concentration and centralization of capital cannot be dichotomized from the process of competition (Clifton 1977; Shaikh 1980, 1982; Weeks 1981a; Semmler 1984; Bina 1985, Ch. 6).

The oil crisis of 1973–74 is thus seen to result from the direct political action of OPEC. For instance, Tanzer argues: "As a result of the Arab oil embargo in late 1973, the OPEC countries effectively took over the ownership of their crude oil reserves and oil pricing, while the companies became primarily suppliers of technology and markets" (1980: 110). These changes seem to be at a phenomenal level, even entirely arbitrary in nature, if one does not accept the arguments advanced by the dependency theorists. These analyses are arbitrary insofar as they are not the outcome of an identified mechanism of value formation in production and of eventual price determination via competition. Consequently, most of these theorists do not offer any systematic analysis, except through alleged unequal exchange.

In this category, as in others, the various theorists do not completely agree with each other. For instance, in his explanation of the oil crisis, Tanzer's emphasis is on the monopoly aspect of the international oil industry, whereas Girvan's primary concern is the "OPEC offensive." Finally, some of the dependency theorists allege that the social and political conditions of the post-Vietnam era, that is, the defeat of US imperialism, are the cause of the oil embargo and the "OPEC offensive." Still others put together numerous factual observations, such as the US political decline, increased participation of OPEC internationally, and increased income of the OPEC countries, to demonstrate that dependency theory is compelling. These arguments are controversial and misplaced: partly because the hypothesized "OPEC offensive" supposedly resulted from a contradiction between the masses of the Third World and US imperialism rather than from the increased development of capitalism and transformation of the oil sector in the OPEC nations; and partly because the political actions of OPEC by themselves cannot possibly be understood without a prior analysis of the underlying cause of the crisis. This arbitrary change of focus, combined with the lack of appreciation of a materialist theory in the political economy, shunted many of these well-meaning, left-leaning scholars to a dead end. The "OPEC offensive," far from being the cause of the oil crisis, was simply one of many elements in fulfillment of the decartelization and globalization of oil. Even so, after the passage of nearly four decades of excruciating theoretical debate and empirical findings, Hanieh (2011) speaks of "dependence" and dependency where it comes to Middle East oil. He writes:

> The increased Chinese reliance on Gulf hydrocarbon is matched by the simultaneous *dependence* of the US, Europe and other key states on these same imports. . . . India also *depends* heavily on Middle East oil (pp. 189–190, emphasis in added).

As is demonstrated throughout this book, and despite the frequent allusions to Marx's in Hanieh (2011), such a deduction is a product, that is an exact copy, of an inapt rightwing theory with respect to oil (and global relations).

## Conspiracy View of the Oil Crisis

The conspiracy view of the oil crisis was based on the idea that the US government, in collaboration with the international oil companies and OPEC, deliberately brought on the oil crisis of 1973–74. According to this view, the oil crisis is centered on the rivalry between the United States and Europe and Japan in the context of international trade and balance of payments. It is believed that the sudden increase in the price of crude oil in 1973–74 was the result of the coordinated efforts of the US government, the major oil companies, and the most accommodating members of OPEC, to increase the price of crude oil (Anderson and Whitten 1977, Greider and Smith 1977). Although the United States became a net importer, it is said, the burden of the price hike fell more heavily on Japan and Europe, where almost all the oil consumed was imported (Tsurumi 1975). Another feature of the conspiracy theory is the amount of discretion that it assigns to price determination in oil.

The price hike neither reflects the conditions of supply and demand nor is associated with the restructuring of the industry, but is rather the result of a pure coordinated exercise of political will (Greider and Smith 1977). The plausibility of this rests on the assertion that a great deal of harmony existed between the oil companies and OPEC, and that the primary contradiction (or in the panoramic parlance of traditional left: inter-imperialist rivalry) in the world economy is between the United States on the one hand, and Europe and Japan on the other. The basic error committed by these writers is the exclusive reliance on the balance of trade. Of course, it is apparent that the increase in the price of oil had a relatively more severe impact on the balance of trade of those economies that were heavily dependent on imported oil. But this outcome does not have anything to do with the cause(s) of the oil price change, unless one believes, a priori, that the price of oil is determined by the US government, oil companies, or OPEC monopolistically by discretion.

Once again, we are confronted, more or less, with the same problems that we encountered with the previous theoretical formulations. But here the difficulty has a different dimension. The static and direct determination of prices through monopolistic power is one thing; their conspiratorial determination by a state or an agency is another. This argument offers no theoretical economic explanation of the process of price formation in general, and the formation of the oil prices during the crisis of 1973–74 in particular. Similarly, there is no economic and political connection between the actual cause of the oil crisis and its consequences. There is only the alleged motivation for engagement in conspiracy on the part of the United States against its so-called allies of wishing to have a more favorable balance of trade. Meanwhile, it has become evident that, in a fairly short time after the crisis, that US balance of trade should decline substantially as the favorable differential impact of the oil price hike would have subsided (see Bina 1985: 8–9, Tables 2 and 3).

Another major obstacle associated with this view is the impossibility of its empirical verification. It is practically impossible to prove or disprove that it was through a conspiracy that this crisis was created. Even if one were able to document that in fact US officials have eagerly welcomed the oil price hike, one still has to build a crisis theory independently of such a motivation. Economic crises are not phenomena that can be created or prevented in the course of discretionary actions by one or another authority. That is why the conspiracy theory of the oil crisis is idiosyncratically tautological at its core. The concern of this theory is the state of minds of the US officials, international oil executives, and that of OPEC officials vis-à-vis deliberate, fully calculated outcome. This is an idealism of the absurd, particularly in the absence of solid evidence. This view is pregnant with paranoia; besides, this, in methodological jargon, is a reliance on the *post hoc, ergo propter hoc* fallacy—reading a causal relation into the concurring events. Hence conspiracy theory does not take into account with an open mind the objective economic and political forces that had led to the oil crisis (for a crafty example with bells and whistles see Nitzan and Bichler 1995 and Bichler and Nitzan 1996).

This view, therefore, addresses *effects* of the crisis. And once these effects are demonstrated, these writers will resort to speculation to determine the cause of the change in the price of oil. In this instance, the above theory should be called a "speculative" theory of the oil crisis. It should be recognized that we do not wish to deny the possibility of conspiracy that may have accompanied the process of oil crisis. But in fact, the proof of the existence of such conspiracies depends fundamentally upon the identification of the actual cause of the oil crisis and not the other way around. To deny the validity of the conspiracy theory, therefore, is not to deny the rivalry among the modern industrialized nations, for rivalry is a real occurrence and an objective process in everyday life. The lack of validity of the above analysis is rather attributable to mistaking the reality based on its skin-deep impression. And in so doing, the cause of the oil crisis was reduced to the ad hoc purpose of individual actors by way of conspiracy (Bina 1985, Ch. 2).

In sum, we have seen that all the above theories of the oil crisis are more or less subjective in nature and speculative to various degrees. These theories deal with the effects of the oil crisis, and either partly or entirely tend to regard these effects as causes. Some of the theorists associated with the traditional theories of the oil crisis in fact deny that the crisis was an objective process. Instead, they emphasize changed perceptions. Others in this category, while acknowledging some objective changes, do not penetrate below the surface of appearance far enough and therefore end up describing the phenomenal form of the crisis. The dependency theory of the oil crisis emphasizes the notion of an "OPEC offensive" within "center-periphery" economic relations (Girvan 1975, 1976; Tanzer 1980). Thus, most of the theorists of this category stress unequal exchange and the "challenge" of OPEC as a cause of the crisis. Finally, conspiracy theorists reason that through a demonstration of the differential impact of the oil crisis on the balance of trade of the United States, Western Europe, and Japan, there was a conspiracy on the part of the US government, oil companies, and OPEC in bringing about the oil crisis (see Anderson and Whitten 1977). All these theorists in due course failed to realize that their arguments liken the crisis to a voluntary act rather than an objective social and economic process aimed at restructuring the industry. Thus, they commit the error of taking the effect of the crisis for its cause.

## THE US DOMESTIC OIL

We maintain that the oil crisis of 1973–74 was the consequence of a significant increase in the long-term production costs of oil in the United States prior to October 1973. Due to the nature of the industry, and to its social relations of production, the US oil region has become the least productive in the world. In addition, as a result of the integration of production at the global level, oil values, oil rents, and market prices are no longer subject simply to the framework of national economies, but are determined internationally. A glimpse of capital investment costs necessary to produce a new daily barrel of oil in a selected number of oil-producing countries during the period between1960 and 1972 can be seen in Wyant (1979: 117, Tables 5–17) and Bina (1985: 67, Table 10). It is crystal clear that the US oil production is by far the most expensive in the world; on the other hand, the cost in the Middle East oil region is the least expensive. Thus, our analysis will have to focus on the US oil industry vis-à-vis the Middle Eastern oil industry.

We shall demonstrate that the combination of critical arguments advanced so far, along with the empirical expositions presented below, will corroborate the above hypothesis and support our theory of the oil crisis. The task here is to show that the retention of the aged US oil fields, which resulted in a decline in average oil recovery, was the underlying factor that necessitated the reorganization of production, and formation of new structures in value, market price, and rent (royalties) in the international petroleum industry. This reorganization ultimately emerged through the crisis of 1973–74 which, as we have seen, did not confine itself to one country or two, but quickly swept throughout the entire global structure.

## Diversity of Oil Production

According to the literature on the petroleum industry there are three different ways by which additional reserves of petroleum can be developed and brought up to the surface: new discoveries, extending the old discoveries, and recovering additional oil from the existing oil fields. It goes without saying that these methods differ in the conditions of capital investment and the resultant capital intensity. While useful in principle, statistics provided in this study concerning the classification of domestic oil reservoirs do not provide a clear-cut procedure that would separate the latter two types of oil production, that is, the extension of oil discovery as opposed to the enhancement of oil recovery. But, fortunately, we need not be too concerned about this, since there is a general consensus that these deeper oil wells that produce more than 32 barrels a day are normally the likely candidates for extension as compared with more shallow oil wells of lower productivity. It is the latter category of oil wells that usually becomes the subject of enhanced oil recovery methods through intensification of capital investment. In addition, the sphere of new oil discovery was more productive, both prior to and during the period of the 1973–74 crisis (US Department of the Interior 1967, 1976; US Department of Energy 1978a, 1978b). The significance of the above classification will become clear as we proceed to analyze the impact of long-run investments on the production of oil in the United States during the period leading up to the crisis.

## Intensification and Reversal of the Oil Recovery

More than 90 percent of US crude oil was produced from reservoirs located in nine states during the period 1965–74. Of this, nearly two-thirds was produced in Texas and Louisiana. The remaining one-third came from California, Oklahoma, Wyoming, New Mexico, Alaska, Kansas, and Mississippi. Studying the trend of average oil recovery in the United States and in the principal oil-producing states, in conjunction with the long-run investment per barrel, is one possible way of examining the actual emerging conditions that led to the formation of the present cost structure in the oil industry. The identification of decline, either in the trend of oil recovery from the old oil fields or in the rate of new oil discoveries, in conjunction with the long-term cost of oil associated with these spheres of production during the period of 1971–74, reveals which area actually regulates the restructured industry's market value and prices.

A comparison of these trends requires calculation of the average oil production per well for the above-mentioned major US oil-producing states for 1965, 1971, and 1974. But the average of these averages does not represent the true average, without assigning appropriate weights, such as the actual production shares of these individual states. The comparison of the average oil recovery and corresponding shares of oil production for the US oil-producing states for the period of 1965–74 can be found in Bina (1985: 70–71, Tables 11–13). These findings show that the conditions of capital investments and production during 1971–74 were entirely different from those in the previous period. The increasing volume of capital investments on the existing oil fields, and the production of reserves by way of intensification, extension, and enhanced recovery methods, led to the subsequent decline of average oil recovery per well in the aging US oil fields.[5]

One has to bear in mind that this decline is not simply incidental to the oil crisis of 1973–74. The crisis is, in fact, both the symptom of and the social mechanism for the generalization of production conditions under these newly emerged circumstances. The magnitudes of the cost of production and of the individual value produced for the aging US oil fields, due to the intensification of production, had significantly increased. Being the least productive of all US oil fields, their corresponding newly formed individual value has become the social value of the entire industry.

## Costs and Individual Values

In this section it shall be argued that total US oil capital expenditure per barrel tripled during the period of 1971–74. During the same period, the US domestic oil price also tripled. In this particular analysis, the period of 1971–74 provides the framework for understanding the oil crisis of 1973–74. Empirically, the relationship between the total US cost of investment (per barrel) and the average US wellhead price of oil is the basis for the argument that the increase in the price of oil was greatly influenced by an increase in its capital cost. Given the fact that the US region was the least productive oil region of the world, one might have expected that, with the internationalization of production, the global price of oil would be determined by the newly emerged US cost structure. In other words, the US production conditions determined the regulating market value of oil for the entire international oil industry. To appreciate the source of this cost increase, one has to split the effect of exploration from that of the development capital expenditures. These are the two most significant constituent parts of long-run capital costs.

Hence, the primary source of cost increase has been the US oil development costs. These costs include those capital expenditures that were applied to the aging US oil fields, known as "extensions" and "revisions" in the literature. This is contrary to popular belief that in search of new oil discoveries the US domestic oil encountered an increase in the cost of oil. Moreover, the highest increase in costs had occurred in the old oil fields, from $1.075 to $3.885, more than 300 percent (per barrel), in the 1971–74 period—right on the cusp of the crisis. On the other hand, an increase in the exploration costs (per barrel) was only 8 percent, insignificant by comparison. By comparison to the period of 1966–70, this alone led to more than a tripling of cost (per barrel) during the 1971–74 period (see Bina 1985: 74–75, Tables 14–15). This picture is also contrary to the mainstream economic theory that relies (as David Ricardo did) on the extension to marginal lands or oil fields (see next chapter for further details).

Thus, a comparison of the changes that had taken place during the period of 1971–74 with the changes in the average costs associated with the periods of 1966–71 and 1972–75 demonstrates that there was a significant increase in the development costs and total costs per barrel in the latter period. In fact, both the development costs per barrel and total costs per barrel more than tripled during the period 1971–74. This result is in conformity with the tripled level of the US oil prices during the same period. This more than threefold increase in the trend of development costs (per barrel) also conforms to the fact that the oil discovery rate (i.e., fresh exploration) of nearly 15 percent was the consequence of 8.4 percent increase in exploration costs during the period of 1971–74 (Bina 1985: 76, Table 16). This again points to the aging US oil fields that were scattered in the lower 48 states and thus were the ground zero of global restructuring of oil.

The correspondence between the fourfold increase in the "posted price" price of oil in the Middle East, Africa, and Latin America, on the one hand, and the more than threefold increase in US oil development costs, on the other, reveals a significant connection among all oil-producing regions that has been objectified, and at this juncture, through crisis, paved the way for the unification and globalization of oil. The primary requirement for the globalization of oil, of course, was also dependent upon the breaking free of the Middle Eastern (and African and Latin American) oil from the jaws of the antiquated, semicolonial oil concessions under the now defunct IPC; hence the antagonism by the oil-producing countries and insistence by the cartel in stretching the limits of the semicolonial oil order. The point of contention was primarily the *oil rent*, albeit beneath the guise of "equity participation," joint ventures, and so on, which was eventually increased based on differential productivity of oil across the globe. Through the objective force of the crisis the status of rent in oil had changed overnight from a bone thrown to the oil-exporting countries to a full-fledged, respectable category in the annals of political economy. Accordingly, the US oil region—being among the least productive in the world—had to be restructured by the force of the crisis and by the absence of the cartel (i.e., presence of competition) that kept lifting oil from the aged US domestic oil fields. The higher market price based on the regulating capital of the least productive oil fields thus synchronized the globalization of crude oil through worldwide competition.

At the onset of crisis, there appeared in the US oil region two distinct possibilities either (1) abandon the majority of the aging (and declining) US domestic oil fields for the sake of existing price structure, or (2) allow them to set the threshold for the least productive oil region in relation to all other oil regions of the world. The first alternative, of course, proved impossible due to the fact that the global demand could not have been fulfilled without full-blown production from these oil fields, in which case a worldwide shortage could have led to increased prices and thus pressure (as well as incentive) for those oil fields to reopen and produce again. In other words, the result would have inevitably been the same. That is why with emergence of new value and new market price such possibilities turned rather inevitably into actuality during the oil crisis of 1973–74. This is the kernel of an *objective* theory of crisis articulated in our earlier work and in this book.[6]

It might be asked why, in the event of such an unprecedented increase in the US production price during the oil crisis of 1973–74, these highly unproductive and seemingly inefficient US domestic producers were not entirely eliminated from the market by the more efficient producers of the Middle East. First, the significance and the size of the US oil production, both prior to and during the period of 1973–74, should not be underestimated. We know that US oil production has been traditionally the largest among the oil-producing countries. Given the structure of social demand and the existence of shortages, it would seem unlikely that the aging US oil fields could have been forced out of the market during the oil crisis of 1973–74.

Second, the productivity of the US oil industry varied substantially from field to field, and from oil well to oil well. In fact, there is a great deal of productivity differentiation even within the continental United States. Thus, even if some of the most unproductive oil fields might have been eliminated, the bulk of US oil production originating from the more productive areas would have remained intact, the relatively less productive fields located in the North Sea and elsewhere in the world being squeezed out of the market first. A remarkable example that supports the above point is the wholesale elimination of the least productive US oil fields during the oil crisis in 1986. The prolonged condition of oversupply that had been created principally by Saudi Arabia's overproduction, and that has, once again, resulted in the restructuring of global production in the oil industry, seems to be the leading cause of this massive elimination of the high-cost producers within the US. This, of course, was not despite the US foreign policy in the region but because of it. The Saudi's move, while economically self-injurious and irrational, was probably motivated by bolstering Saddam Hussein's position against Iran in the ongoing Iran-Iraq war, which was parallel with the de facto US policy by the Reagan administration, namely, nipping the new Iranian regime in the bud when the conditions were still in flux. Even though the production price is still being originated from the US oil region, the regulating value of the entire industry today conforms to the individual market value of the more productive US oil fields. As a result, we may observe a substantial decline in the magnitude of market prices that are necessarily gravitating around the center of a newly formed regulating value or the price of production.

Finally, the notion of "political intervention" and the immediacy of political motivation should not be overemphasized. The *polity* is a distinct, objective, and historically specific social category. We admit that the nature of the relationship between the political and the economic realm is somewhat controversial, but despite the dialectical interactions of the two, one cannot rely on the immediacy of these interactions. In other words, the study of political motivations (as the end results) may not necessarily bring us to the understanding of the social forces that may or may not have anything to do with those motivations. The social significance of polity, in our opinion, is much more than that. Kolm argues: "To say that the economy, in a more traditional sense, is inextricably involved *with* other social phenomena, e.g., psycho-social, 'pure' political, etc., is to state a self-evident truth. It is, in fact, [must be considered as] an aspect, a side of the same social whole" (1981: 210). It is in this particular sense that political motivations (e.g., the state of mind of politicians, such as conspiracy) cannot be used to identify, or even to justify, the fundamental basis of their so-called originating process, except by way of speculation. The task, however, is to understand the process directly, and then possibly to find out whether or not there may have been a conspiracy involved.

## The New Oil Discoveries

As we have pointed out, the production of oil reserves assumes three different forms. We have dealt with two of these possible forms, that is, with extensions and revisions, in the previous sections. It is the task of this section to deal with the remaining type, new oil discovery. First of all, we need to follow the same pattern of periodization that was utilized in the previous section, for the sake of uniformity and the possibility of comparison. Thus, the trend of productivity of the US new oil discoveries for the familiar periods of 1965–71 and 1971–74 has to be analyzed to determine whether or not they correspond to the crisis period.

The productivity of new oil discovery is usually shown by the number of newly discovered barrels of oil per foot of drilling during a specific year. To be able to come up with the corresponding figures for oil exclusively for each year, however, we must separate the volume of oil exploration from that of natural gas and other associated hydrocarbons. Bina (1985: 76, Tables 16) points to a 15 percent improvement in the rate of discovery for the period of 1972–75 over the period of 1965–71. Clearly, the above improvement, if it were determining of value and of price, would have implied the existence of a market price of even smaller magnitude; this is exactly the opposite of what actually took place in the crisis.

During the periods of 1965–71 and 1971–74, the corresponding trends of productivity of new oil exploration exhibit respectively a decline of 55 percent and an increase of 113 percent. Even if one disregards the extremely low productivity of 1971 and considers the periods of 1965–72 and 1972–74 for the analysis, one still will come up with 23 percent decline and 24 percent increase, respectively. In addition, the corresponding average oil-finding ratios for the periods of 1965–72 and 1972–74 are 14.27 barrels per foot and 15.39 barrels per foot, somewhat of an improvement. As for the total finding rates (oil and gas), for the periods of 1966–71 and 1971–74 one finds a similar trend. More specifically, we have a 43 percent decline in the productivity of hydrocarbon exploration in the precrisis period and a nearly 39 percent increase in the period leading up to the crisis. Bina (1985:78, Table 17) shows the existing trends of productivity within the sphere of new exploration for both the precrisis period and the period immediately leading up to the crisis.

What is noteworthy here is that during the period that coincides with the oil crisis of 1973–74, productivity of both new oil and total hydrocarbon exploration increased substantially. This indicates that if the oil crisis were the result of structural changes in the sphere of exploration, we would have had a decrease rather than an increase in the productivity trend of this sphere of production, whereas the productivity of exploration exhibited a significant decline during the precrisis period. But the decline in the productivity of exploration during the period of 1965–71 seemingly did not have an impact upon the value and the price structure of that period. The above conclusion is also true for the period of 1971–74. In other words, the conditions of capital investment and the impact of productivity developed in the sphere of exploration do not seem to conform to the circumstances that led to the oil crisis of 1973–74; it would be reasonable to conclude then that the center of the oil crisis was actually within the aging and already producing oil fields.

### Fragmentation of Leases and the US Productivity Decline

As shall be explicated in the next chapter on rent, there has been a significant distortion in the exploration of oil in the lower 48 states with respect to the size of the lease-hold due to "divided ownership of land in the United States" (Miller 1973: 415). This situation had given rise to the reduction of benefits to the original discoverer of the oil fields and took center stage as noticeable evidence that is rarely noticed in the period leading up to the crisis (see Bina 1985, Ch. 5). Thus, the pattern of the fragmentation of land ownership with respect to the system of leasing for oil afresh can be a part of the reasons why there was an aversion by the independent prospectors in the lower 48 states in the United States to investment in the larger oil fields in need of assembling sizeable tracts of land. This goes for the fragmentation of leases with respect to secondary and tertiary recovery, and for the unitization of the oil fields according to a "predetermined" technical schedule. This brings us to the fundamental issue that the intensification of capital investment within existing oil fields was clearly the consequence of the impediment of the prevailing pattern of land and lease ownership in US oil production. In this context, the structure of landed property and the fragmentation of oil leases played an influential role in the direction of capital investments and the structure of accumulation in the US oil industry long before the oil crisis of 1973–74, and thereby set a new basis for the formation of values, oil rents, and market prices across the globe.

Let us now switch gears to briefly look at the state of US oil recovery prior to the crisis. The trend of the average level of oil recovery would paint the picture of the evolving production conditions that had been prevailing in the US oil fields. Given the geological structure of the oil reservoirs, production conditions are the consequences of the circumstances of capital investment in the entire industry. The concentration of capital investments in the aging oil fields, if widespread, can require the restructuring of the entire industry. In other words, if the cost structure of production in the aging oil fields, as a result of further "capital deepening," were to rise, then the magnitude of the individual value associated with those oil fields and possibly of the social value for the entire industry would have to change as well.

It turns out that the trend of average US oil recovery, that is, the trend of the weighted average of daily barrels of oil per well for the period, 1965–71, shows a considerable increase. But the recovery trend for the period, 1971–74, the period leading up to the crisis, shows exactly the opposite. Considering that investments are made either in the form of "capital widening" or by way of "capital deepening"—the extension of the existing oil fields, or the massive application of enhanced recovery methods—these figures ineludibly imply capital investments through "capital deepening."

In addition to the above conditions, the average life span of US oil wells was substantially shortened during the period, 1971–74, as compared with the trend of 1965–71 (American Petroleum Institute 1979). This shows that the depletion of existing oil fields increased substantially as further capital investments were applied to the existing oil reservoirs on a massive scale during the period of 1971–74. Likewise, the rate of cumulative abandonment of oil wells during the period of 1971–74 declined as well. Abandonment is considered when an oil well is commercially exhausted. The average rate of oil-well abandonment per year for 1971–74 was 1.54 percent or about half of the corresponding figure for the period of 1965–71. As a result, one has to conclude that in the period of 1971–74 a substantial number of oil wells, which in previous periods would have been abandoned, were retained for further production. And to keep the production flowing from such wells, additional investments of capital were necessary. Given the conditions of the oil wells, one has to expect an eventual decline in the productivity of such investments.

As we have pointed out earlier, the oil crisis of 1973–74 is the consequence of the generalization of the conditions of production in the aging US oil fields for the entire industry. At the same time we have shown that, due to the existence of a particular property relation within the oil industry, the least productive producer tends to determine the magnitude of value for the entire industry. Consequently, the more productive regions, in addition to their share of normal profit, can appropriate a surplus profit in the form of oil rents or royalties (Bina 1985). Finally, the generalization of production conditions in the least productive US oil fields led to the formation of a new value and cost structure, oil rent, and of newly formed market prices within the context of the entire industry globally.

## Alternative Theory of the Oil Crisis

To develop an alternative theory of the oil crisis, one needs to start from the critique of the prevailing views on the oil crisis of 1973–74. As we have seen earlier, the majority of explanations tended to explain the events that occurred during the period leading up to the oil embargo. But none of these theories was able to show the relationship between these outward appearances and the inner essence of the oil crisis. What is lacking in these explanations is the identification of the law of motion of the crisis. It would appear that they all missed the underlying cause of the crisis because they took the effects of the crisis at their face value. Thus, the existing theories of the oil crisis, from those with the most conservative to those with the most radical orientation, suffer from the dominance of detached idealism and the lack of proper perspective.

An adequate alternative theory of the oil crisis should reflect the internal development of the international oil industry, and the underlying forces that led to the crisis. This requires the explanation of the timing of the oil crisis. We have seen that none of the existing theories was able to demonstrate, for instance, why oil prices changed so drastically, except by way of the neoclassical theory of monopoly or cartel. But the underlying forces of monopoly or cartel during the time of the crisis were not adequately identified, except by pointing to the "dependence" of the United States on the foreign oil. Thus, the static notions of monopoly and cartel, combined with that of subjective determination of prices by the US government, were among the reasons for the oil crisis given by most writers.

## Periodization of the Middle Eastern Oil Industry

The history of oil production in the Middle East has passed through three distinct periods: (1) the era of colonial/semicolonial oil concessions (1901–50), (2) the transitional stage to capitalist development (1950–72), and (3) the post-crisis era of modern capitalist production (Bina 1985, Ch. 3). The era of colonial/semicolonial concessions is the period within which the international oil companies divided the entire Middle East region among themselves by obtaining concessionary rights for exploration and production of crude oil. The areas subjected to these concessions were often almost the size of the countries involved, with contracts that lasted nearly up to a century (Cattan 1967a). The existence of precapitalist social relations, along with the lack of private property in land and the political dominance of international capital, were the principal features of this period. The second period is associated with the further development of capitalist production in the Middle Eastern oil industry. The distinguishing features of this period are nationalization of oil in Iran (and its reversal through the 1953 CIA *coup d'état),* and the subsequent de-nationalization and formation of OPEC (Walden 1962, Rouhani 1971, Mikdashi 1972). With the development of capitalist social relations under the economic and political dominance of international capital in this period, the process of the internationalization of capital in the oil industry accelerated. Finally, with the commencement of the third period, value formation, and with it price formation in the petroleum industry, for the first time in its history, the industry took on a modern international dimension that extended into the entire global structure (Alnasrawi 1985, Bina 1985).

## Property Relations and the Oil Rent

One of the distinguishing features of oil production is production through the intervention of the ownership of oil fields. In other words, production of oil in capitalism is dependent upon the surrender of rights of ownership of crude oil to the capitalist producers by the owners of the subsoil. It is the stipulation of such surrender that sets the limits to intervention of the landed property in petroleum production. This intervention in the process of production through competition leads to the phenomenon of oil rent, just as the intervention of landed property in agricultural production leads to agricultural rent.

Historically, oil rent, like any other social phenomenon, has passed through different and specific stages of development. During the era of early concessions, when the domination and subsequent expansion of international capital in the Middle East and elsewhere in the Third World had just begun, a rudimentary and still undeveloped form of oil property relation emerged that was associated with the state ownership of land in most societies in these regions. Because oil rent at this particular stage was not a barrier to production, there was no impediment to capital investment in oil production. Therefore, in the early period, the determining condition of the magnitude of oil rents was based exclusively upon the direct political dominance of international oil companies within the entire Middle Eastern oil region. This is also true of other regions as well. With the gradual development of capitalism, the emerging oil property relations that led to the phenomenal form of rent, in conjunction with transnationalization of capital, have potentially placed all the oil-producing regions in an organic relation with one another. As the international oil industry extended its range of economic and political domination over the entire globe, the extent and intensity of this organic relation increased. Meanwhile, one has to realize that production of oil was taking place under very different technical and social production conditions in the various oil-producing regions of the world. Thus, the magnitudes of production costs and individual values produced in these regions must necessarily be different from each other. With the social value for the entire international industry emerging from the least productive region of the world, production from the more productive oil regions will accompany differential oil rents over and above the general rate of profit.

## Value Formation and Price Determination

It should be pointed out that there are many types of prices in the oil industry, with different and distinct characteristics. To name a few, there are "posted prices," "spot prices," and "buyback prices." From the standpoint of political economy, however, one has to know the difference between the "oil spot prices" and "posted prices" and the way in which they relate to the process of value formation in the oil industry.

The "spot price" is the price of a single volume of crude oil bought and sold in the spot market (Minard 1980). This is the price that is usually referred to as the market price in economics texts. A characteristic of the "spot price" is that it fluctuates around the center of gravity of social values in the entire industry. Such fluctuations are brought about by market conditions in which demand and supply play an important part.

The "posted price," on the other hand, is a mechanism that was traditionally used for the internal transfer of crude oil within the network of major oil companies and their subsidiaries. The major significance of "posted price" is in its usage in the calculation of oil rents (royalties) since the 1950s (Cattan 1967a; Bina 1985, Ch. 5). After the replacement of the early oil concessions by the new method of 50–50 profit sharing, the "posted price" was used to determine the share of the oil revenues due to the oil-exporting governments. Historically, this represented a further step in the development of modern oil rent in the Middle Eastern oil industry. Although the "posted price" does not function as a price in an ordinary sense, it nevertheless has been constituted as a variable basis for the determination of oil rents since the transition from the era of early concessions.

Even though the meaning of the term "posted price" has not changed since its emergence in the 1950s, its essential characteristic, as a mechanism in determination of the oil rents, has nevertheless evolved. This change is associated with the gradual increase in the level of socialization of production, the internationalization of capital, and the development of full-fledged capitalist relations globally. One of the significant events during the posttransitional period was the emergence of spot markets in crude oil that initially aided the Iranian, Nigerian, and Libyan governments to receive unprecedented bids of $17 to $24 per barrel during the month of December 1973 (Penrose 1975: 51). As we have observed, the spot price or market price did not decline significantly after the removal of the embargo, even with the world economy moving into a major recession during the period of 1974–75. During the embargo period the "posted price" increased fourfold to the level of $11.65 for the marker crude (Arabian light crude oil of 34-degree API gravity) (Blair 1976). This change increased the magnitude of differential oil rents not only for the OPEC nations but also for other oil-producing governments, such as the United Kingdom and Mexico. Almost simultaneously, the wellhead price of the so-called US new oil increased to the world level.

Since 1970, the international oil industry has become an organic whole, with all different oil regions of the world in mutual relationship. The oil crisis of 1973–74 is therefore historically the first economic crisis of the industry at its present (global) stage. Although the above crisis was, for obvious reasons, associated with oil shortage and the consequent supply restrictions, one cannot explain it through oil shortage alone as was often previously the case.

As pointed out earlier, the magnitude of value is regulated by costs and conditions of production in the US domestic oil fields, since the latter is the least productive oil region. As the total costs and development costs of oil (per barrel) had almost tripled (and the trend of development costs alone explains that), this center of gravity had to rise significantly. This increase is the reflection of the increased level of expenditures that are made to develop oil from the older oil fields within the US region alone.

The US oil region has been the least productive area even prior to the threefold cost increase of the period of 1971–73. But continuous intensification of capital investments on the existing oil fields, in conjunction with the fragmentation of the existing oil leases, led to even further increases in the magnitude of US production costs. These changes would necessarily require comprehensive reorganization with respect to the structure of oil production, commensurate with the changes that materialized through the crisis (Bina 1985). Therefore, the Arab-Israeli war of 1973 was the triggering point, not the cause of the crisis (of restructuring). During this reorganization, in conjunction with a substantially higher magnitude for regulating social value, increased levels of differential oil rents, and market prices emerged globally.

As for the 1986 oil crisis, there is a substantial decline in spot market prices together with a great deal of bankruptcies (plugging of high-cost oil wells), which in turn have led to speedy elimination of relatively unproductive fields in the continental United States, a significant number of which are located in Texas, Louisiana, and Oklahoma. The severity and suddenness of these bankruptcies are clear signs of a restructuring of oil production and the formation of a substantially lower regulating market value (and production price) for the entire global oil industry. The so-called free fall of the oil prices, as indicated by most market analysts, is not a pure market phenomenon but rather a market response to abrupt structural changes that were under way. Given the political anxiety over falling oil prices in the mid-1980s, despite the currency of market ideology, the Reagan administration leaned on Saudi Arabia to put a lid on oil production—albeit, after letting the Saudis flood the market to destabilize Iran's economy during the war with Saddam Hussein. Yet, it was not possible for the United States or any other entity in the world to reverse the course of history and to circumvent the epochal transformation that had already permeated the oil industry's global structure.

## Monopoly and Competition in the Oil Industry

The existence of the dominant but erroneous concept of monopoly is common to all the prevailing views on the oil crisis. Without exception, all the existing theories, either implicitly or explicitly, tend to agree that the price of oil is directly determined by the OPEC cartel, or through the monopoly of oil firms, or both. For instance, the dependency theory of the oil crisis implies that oil prices prior to the events of the early 1970s were determined by the oil companies, but after the "OPEC offensive," which was a move against unequal exchange at the international level, they were set by the oil-exporting countries. The conspiracy theory argues that the increase in the price of oil was primarily initiated by the US government in conjunction with the major oil firms and the concerted effort of OPEC. The conventional theory of the oil crisis contends that controls of the production, distribution, and marketing operations do not allow the laws of supply and demand to operate properly. Therefore competition is imperfect (Salant 1982, Robinson 1969, among others) for the scholars who invest their model on the illusory and idealized notion of "pure competition" to present a realistic picture, having little grasp of the fact that the reality is what there is—it is neither perfect nor imperfect (see Bina 2013, Weeks 2013). The common denominator of all these views is the quantitative theory of competition and monopoly (see Weeks 1981a: 152–58). The notion of monopoly is perceived to be dependent upon the number of firms within the industry. Accordingly, if the number of firms operating in an industry is small, it is called a "monopoly" or an "oligopoly." On the other hand, if the numbers of firms in an industry are many, it is believed that competition prevails (see Clifton 1977; Shaikh 1980, 1982; Semmler 1984; Weeks 1981a, 2013; Bina 1985, 1989a, 1989b, 2006, 2013 for a critique).

In contrast with the above views, since competition and centralization of capital (i.e., monopoly) are not mutually exclusive in the process of the production of value, one cannot talk about monopoly without competition. The formation of value in an industry in general, and in the oil industry in particular, necessarily emerges through competition. In the oil industry, just like in any other industry, competition among different production units, with different individual values, leads to the formation of social value for the entire industry. In addition, production of oil is intertwined with the formation of rent, which in turn develops through competition (Fine 1983, Bina 1985). Thus, competition as an inner nature of capital will always be present in the process of accumulation and value formation. Accordingly, the competitive struggle among capitals leads to concentration, centralization, and the further integration of capital in the accumulation process; this competition leads to integration and further integration leads to further competition. As a result, it is hardly surprising that the existing theories all failed to recognize the true nature of the oil crisis of 1973–74, and the significance of postcrisis developments.

## Concluding Remarks

We attempted to develop an alternative to the prevailing views concerning the oil crisis of 1973–74 by demonstrating that the above crisis was qualitatively different from the simple international shortages that occurred in the past. We contended that the above crisis was the consequence of the globalization of production in the oil industry, in conjunction with the reorganization of US production that has brought all the oil-producing regions into an organic whole, leading to a unified system of value, price, and rent determination through competition.

To be able to explicate the 1973–74 oil crisis one has to theorize it first. In doing so, we have developed a theory of globalization of the oil, with distinct characteristics as differential oil rents and competitive prices, formed in the spot and futures markets, all based on Marx's theory of value in capitalism proper. The globalization of the oil industry in the Middle East has gone through three distinct historical stages of development as specified above. These stages are the pieces of the puzzle of evolutionary change toward the globalization of oil. With the gradual development of capitalist social relations in the Middle East and elsewhere in the world, the aggregate embrace and embedment of petroleum, as an essential and trailblazing sector of the world economy, is to state the obvious. What is not so obvious is how to reach a theoretical and empirical understanding that, despite the daily contingencies and contradictory events, enable us to decipher the secret of our epoch through the investigation of this particular medium—globalized oil.

We explained, following the crisis, how the characteristic and conditions of the US oil production (world's least productive oil region) turned up to regulate the magnitude of production price (or value) of oil throughout the world. A substantial rise in the investment costs of US oil—combined with the highly fragmented oil leases—under the purview of the Achnacarry (1928–72) during the period leading up to the crisis was a critical part of this puzzle. The necessity of restructuring and rationalization of the US oil fields was as important as the demand for doing away with the antiquated semicolonial oil concessions in the Middle East, Africa, and Latin America. But neither that *necessity* nor this *demand* was on the agendas of the United States and the IPC, one the one hand, and the rulers of Iran, Saudi Arabia, Kuwait, and other little Sheikdoms in the Persian Gulf, on the other. The latter party just wanted "equity participation" and protection. The former party desired a high-handed foreign policy (via the Nixon Doctrine), and cartelized control over the oil. But the authenticity of change is not a matter of choosing, as history is replete with the carcasses of choice in these important matters.

Globalization of the oil industry was not a simple oil shock as perceived by the popular sentiment; it was an upsurge that shifted the tectonic plates of the polity and a harbinger that prepared the world for the eventuality of the fall of Pax Americana by the end of the very same decade.

# 2

# World of Modern Petroleum and the Oil Rent

*The landlord demands a rent even for unimproved land, and the supposed interest or profit upon the expense of improvement is generally an addition to this original rent.*

—Adam Smith
*The Wealth of Nations* (1977 [1776]: 248)

## Introduction

In this chapter, the objective is to unravel the enigma of rent and to present a well-defined theory of rent unique to the global oil industry. The truism of the above citation from Smith's magnum opus shall be a piece of the puzzle that will hopefully find a specific resolution for the oil industry—albeit without the entrapment of mainstream economics monopoly—in this chapter. This theory must patently originate from the valorization of the landed (surface and/or subsurface) ownership—which in our case extends to the ownership of oil deposits—toward the exploration, development, and the production of oil. Valorization of the landed property is an act of capital as a modern *social relation* (not an inanimate thing) through the *medium* of nature. This mediation is socially involuntary (i.e., objective) and has no bearing on the wishes of the individual owner of the land or the deposits underneath. The oil deposits once valorized step into a mutuality of synthetic relation with capital, and just like *space* and *time* directing space-time in modern astrophysics, do not have an independent existence from one another. In other words, there are no separate logics, one for property in nature (geography or territory) and another for capital where it comes to the notion of rent.[1] This would bring to light the missteps in the theories that are either hinging on the game-theoretic approaches, say, for OPEC oil rents, or rather imprecisely placing rent on the same footing as simple taxation by the government, or, even more outrageously romanticizing the war of capitalist competition as a bundle of harmony and then proceeding to vaccinate it with a bit of realism as "imperfect competition"—thus *rent* for alleged imperfection. *Valorization* in modern political economy may well be considered as equivalent to *space-time* in modern cosmology. This is essentially what the theory of value does in modern, mature, and highly integrated capitalism. Similarly, this is the kernel of our materialist (i.e., scientific) methodology toward a theory of oil rent—as it was in Karl Marx's for agriculture well over a century ago. And this is what our guidelines are when we speak of rent as it relates specifically to the oil (and energy) sector throughout this book.

To be able to advance a unique theory of rent for the oil industry, however, one has to proceed from the fact that rent in modern capitalism is not a general theory. And if there is any, it either exclusively attributes to nature or relies completely on production, irrespective of the mediation of nature. In either of these cases such a theory is no more than an empty shell. Rent is the specific effect of the synthesis of capital and the landed property. And, given the differences of historical conditions of each class and/or type of the landed property (in mutuality with capital), one might find contingencies that differ from industry to industry. Accordingly, there is no one-size-fits-all theory that can fit all rents, thus a unique theory must be put in order for each and every type of rent-collecting industry in our time. That is why any judicious theory of oil rent must be informed of the limits and exactitudes, including globalization, of the oil industry across all regions of the world. (Bina 1985, Ch. 5). Hence, without an intimate knowledge of the industry (necessary conditions), combined with a well-informed historical and theoretical grasp of the political economy of rent (sufficient conditions), speaking of oil rent would remain devoid of foundation and thus have limited application.

Economic or ground rent has a multiplicity of applications that are not adequately dealt with by the dominant orthodoxy, other than imposition of *external* or monopolistic conditions. What is forcing the majority of theorists to speak of rent presently seems to be none other than "practical necessity" that is loosely connected with a consistent theoretical foundation. It is in this connection that we will attempt to evaluate the prevailing conceptions of rent in general, and the phenomenon of oil rent in particular. This of course, among other things, leads us to a brief evaluation of a selected number of rent theories that were developed historically in conjunction with agricultural production. (Ricardo 1976, Ch. 2; Marx 1981, Part VI).

The notion of ground rent in economic literature has its origin long before the emergence of the classical political economy. With the development of classical economics, *ground rent* became an integral component of the economic theory of this period and produced a great deal of controversy whenever it was applied to the economic and social conditions of the time. During this period, from concern over the fluctuation of corn prices, which was initiated by James Anderson (1777, 1797),[2] and the exclusive analysis of Adam Smith (1977, Ch. 11) during the late eighteenth century, to the extensive debates of Ricardo and his contemporaries on the nature and consequence of the *corn laws* during the early nineteenth century in England, the importance of *rent* and the significance of the analysis of landed property in political economy seem indisputable (Ricardo 1976, Chs 2, 24, 32; Malthus 1815; Torrens 1814, Ch. 8; West 1815).

It is no exaggeration to say that the study of political economy by Karl Marx could not have been accomplished to the extent that it was had it not been for his analysis of *ground rent*, as he devoted a considerable portion of his writings to landed property and its ramification for value and prices in modern agriculture (Marx 1968, 1981, Part VI). Toward the end of the nineteenth century the debates over rent in the United States were influenced by Henry George (1938) and his followers, who did not fundamentally depart from the original perspective of their classical predecessors and boldly proposed a system of single tax on the land to be disbursed as the main source of state revenue. In the meantime, with the development of the "Marginalist school" that has gone through decades of debate over the nature of *rent*, this query has still been kept from adequate scrutiny (Brown 1941, Fine 1983). For instance, since the advent and adoption of "general equilibrium," the notion of rent has been either evaporated from the models, or recognized only (and mistakenly) in models that are built on "imperfect completion," oligopoly, and monopoly—that is, a deviation from the fiction of "perfect competition."

## Ricardian Theory of Rent

Ricardo's concept of rent presents itself essentially in a two-prong question of *scarcity* and the lack of *homogeneity* of land. He declares: "If all land had the same properties, if it were unlimited in quantity, and uniform in quality, *no charge could be made for its use,* unless where it possessed peculiar advantages of situation" (Ricardo 1976: 35, emphasis added). Here both the limited quantity and heterogeneity of land are factors that led Ricardo to develop his theory of rent.

He argued that as the margin of cultivation is extended to use land of inferior quality, cultivation on the land of higher quality results in rent. The rent of this type is called differential rent by Ricardo. Moreover, this differential rent reflects the difference between the quality of "marginal" and that of "infra-marginal" land as cultivation is extended. Another way to perceive such a rent is by noting the limited quantity or *scarcity* of land. That is why Marshall, who did not quite abandon his classical orientation, in evaluating Ricardo's conception of rent, pointed out that, "In a sense, all rents are scarcity rents, and all rents are differential rents" (Marshall 1961: 144). The origin of rent, as Ricardo saw it, is due to the physical characteristics of nature, which are perceived to be universal no matter what the *character* of production might be. As a result, this same concept is applicable to all periods of history. In fact, the consequence of such an analysis is that the phenomenon of rent, as it relates to modern landed property, is *not* unique to capitalism.

Ricardo attempted to make a crucial distinction between rent and profit as part of a more general theoretical development, that is, the laws that regulate the realm of distribution of the produce of the earth among the "three classes of community, namely, the proprietor of the land, the owner of the stock or capital necessary for its cultivation, and the laborers by whose industry it is cultivated" (Ricardo 1976: 3). It is not surprising at all, therefore, that the *genesis* of rent in Ricardo is more in tune with physical or *technical* considerations rather than *social necessity*. The cause of rent here is seen in the extension of cultivation instead of the monopolization of nature, and attempts by capital to overcome it, which necessarily leads to the development of a particular structure of property relations in agricultural production; Ricardo was against the landlords and he was cognizant of landlords' political influence through his own practical legislative experience in the early nineteenth-century British Parliament. Nevertheless, as we demonstrate in this chapter, his "natural law" methodology had no room for rent as a specific (not general) category created by the synthesis of capital and landed property.

## MARSHALL'S SCRAMBLED CONCEPT

Alfred Marshall's notion of rent (and quasi-rent) appears to be a midway between the Ricardian theory and the transmuted neoclassical rent in the mainstream economic theory of today. Therefore, a brief overview of Marshall's contribution can be beneficial for the purpose here. Marshall maintained: "While man has no power of creating matter, he creates utilities by putting things into useful form; and the utilities made by him can be increased in supply if there is an increased demand for them: they have a supply price" (Marshall 1961: 144). He went on to say that although land is among the objects of utilities, one cannot simply exercise any control over it due to the fixity of the supply of nature. Therefore, land has no particular price the same way as any other articles of utility. Marshall thus made a clear distinction between land and the products of land, when he stated that "the fundamental attribute of land is its *extension*" (Marshall 1961: 147, emphasis added). It should be noted that the similarity between the Marshallian conception of rent and its counterpart in the Ricardian tradition is not only due to the emphasis on the extension of cultivation. Marshall, like Ricardo, considered the "Inherent properties, which the land derives from nature... It is chiefly from them that the ownership of agricultural land derives its peculiar significance, and the Theory of Rent its special character" (Marshall 1961: 147). However, there are striking dissimilarities between the Marshallian and Ricardian theories of rent: Ricardo, following Smith, attempted to make a distinction between *rent* and *profit*, while Marshall tried to do exactly the opposite. Marshall remarked:

> [T]he full rent of a farm in an old country is made up of three elements: the first being due to the value of the soil as it was made by nature; *the second to improvements made in it by man;* and the third, which is often the most important of all, to the growth of a dense and rich population, and to facilities of communication by public roads, railroads, etc. (Marshall 1961: 156, emphasis added).

First, it is evident from the above passage that the concept of rent is the proximate outcome of nature. Second, rent is necessarily a universal concept and as such does not require to be specifically dealt with through the analysis of modern landed property, unique to capitalism. Third, the contradiction against the expansion of capital imposed by private ownership of land and the very social act of appropriation of nature is taken for granted. Fourth, the growth of mass communication and transportation as the source of one form of rent is heedlessly mixed up with ground rent. Finally, as the above passage indicates, not only is there no clear distinction between *rent* and *profit* in Marshall's as can be found in the works of Smith and Ricardo, but Marshall has often misunderstood the implications of the theoretical contributions of his classical predecessors, even though he did not entirely abandon the classical tradition. Despite these differences, however, the notion of rent formulated by Marshall gives rise to a concept that is not modified by the characteristics and social necessities of different historical epochs. Rent becomes a universal (and transhistorical) concept in Marshall's theory as well as in Ricardo's.

## MARX'S VALUE AND THE CONTEXT OF RENT

The articulation of Karl Marx's theory of rent in agriculture is a direct consequence of his critique of political economy. Marx argues that the concept of capitalist ground rent relates to the social conditions of the development of capital in agriculture. Thus the concept of *modern* rent has to be perceived within the capitalist mode of production alone, rather than as a universal phenomenon that is relevant to all history (Marx 1981, Ch. 37). The basis of capitalist rent in Marx's sense lies in the production of *exchange value*, in contrast with rent in the Middle Ages, which is connected to the physical entity of the product and its *use value*. Thus, the notion of modern rent as opposed to rent in precapitalist production does not have its origin in nature. Marx clearly delineated the difference between the two as follows:

> An incorrect conception of the nature of rent has been handed down to modern times, a conception based on the fact that from the natural economy of the middle ages, completely in contradiction to the conditions of capitalist mordent in kind survives from the Middle Ages, in complete contradiction to the condition of the capitalist mode of production, partly in the tithes paid to the Church and partly as a curiosity in old contracts. The impression is thus given that rent arises not from the price of the agricultural product but rather from its quantity, i.e. not from social relations but from the earth itself (Marx 1981: 923).

According to Marx, to discover the genesis of capitalist ground rent, one needs to look into the relations of use value and exchange value in the production of commodities. It is quite true that without the production of use value there is no possibility of producing exchange value. But, accordingly, to be able to make a *distinction* between the capitalist mode of production and all the other precapitalist social formations one needs to look for differences and not similarities. Moreover, it is only under capitalist social relations of production that fundamentally production of *exchange value* intertwines with the exchange of the commodity *labor power*. In this connection, commodities have two different aspects: (1) value in use and (2) value in exchange. A bottle of wine produced within capitalism and a very similar bottle of wine produced within precapitalist social relations cannot possibly be distinguished in appearance, and from the standpoint of its end use it is not known beforehand under what social relations each one has been produced.

Likewise, if the conditions of production of these two bottles of wine were totally *abstracted* from wine's use value, one has no way of knowing about the respective social institutions from which they actually originated. The resolution of the above problem is of paramount necessity for those institutional approaches that are basing their entire theoretical structure on such a distinction. Thus, to be able to make a clear distinction between a specific period of history, that is, capitalism, and other historical epochs, one has to abstract from the similarities by concentrating on the differences. While the utilization of nature is common to all human societies in history, the act of utilization of nature in capitalist production, as opposed to all the previous modes of production, is not an end in itself. It is rather a *medium* through which the social relations among people are regularly transformed into *fetishized* relations among things. Thus, an interest in the institutional characteristic of a specific historical epoch requires an analysis of social relations (Marx 1977, Ch. 1).

In the same manner, according to Marx, the genesis of capitalist rent is not owing to the physical attributes of nature, for these physical aspects by themselves do not automatically lead to the reality of rent. What gives rise to this notion is rather the constitution of those historically specific social relations that would utilize *nature* in the production of value. The monopolization of a natural sphere, on the one hand, and the expansion of capital into it, on the other, leads to the division of surplus value into rent and profit. Therefore, the above processes, that is, both the monopolization of nature by landed property (or any other type of property relations that would monopolize the condition of labor) and the expansion of capital, do not have their origins in the physical characteristics of nature, since these processes are rather the results of the existing social relations and not their cause in our epoch.

Historically, according to Karl Marx, during the transition from the feudal (natural economy) to the capitalist mode of production, a dual transformation emerged as a result of (1) separation of the peasantry from the land and the transformation of this class into a class of wage laborers, and (2) dissolution of the collective ownership of land (owned by the landlords) and its transformation into monopolized private property. The transformation of nature into private property is initiated by capital itself that, in turn, comes to stand against its own self-expansion and accumulation in the form of rent. Thus, the contemporary notion of rent is itself a direct dialectical consequence of the rule of capital and its social origin has nothing to do with the physical characteristics of nature. The notion of rent under the capitalist mode of production is the phenomenal form, a mere reflection of the established norms of capital that manifest themselves in the material form of nature. The nature of the above simultaneous transformations is carefully detected by Marx, where he states:

> The determination of the market-value of products, including, i.e., also of products of soil, is a social act, albeit a *social act*, even if performed by society unconsciously and unintentionally, and it is *based upon the exchange-value of the product and not on the soil and the differences in its fertility* (Marx 1981: 799, emphasis added).

## Rent in Exhaustible Versus Renewable Resources

The genesis of modern rent created the social conditions under which a capitalist *rentier* class came into existence. These social conditions are the effect of capitalist production, that is, production for profit. This process, as has been seen above, had given rise to certain particularities that Marx identified as ground rent belonging to modern capitalism. We have also pointed out that modern capitalist rent is the direct consequence of private appropriation or monopolization vis-à-vis the expansion of capital, resting on the institution of private property in the means of production specific to capitalist societies. Such an understanding, from the standpoint of Marx's rent theory, is a precondition for any serious analysis that dispenses with the unmediated utilization of nature; thus in consequence, starting off with nature in its naked (raw) *physical form*—that is, in its *exhaustible* versus *renewable* distinction—will inevitably take us back to the Ricardian or, worse, the Physiocratic origins for a good runaround (Gray 1914, Hotelling 1931, Symposium on Exhaustible Resources 1974). Besides, such an unmediated mechanism is a remarkable conduit for furthering what is known as commodity *fetishism* in the mainstream economic theory (see Marx 1977, Ch. 1: 163–77).

Contrary to Ricardo's, Marx's rent theory is not caused by the differential quality of inframarginal and marginal land but by the manifestation of a specific social relation. The quality of "marginal land," in association with the assumption of zero rent advanced by Ricardo, is not necessarily in sync with the institutional origin of rent, because Ricardo tends to assume away the historical barrier of rent and the process of overcoming this barrier by capital. For the classical political economists *capital* was not an epoch-bound (specific) social relation but a transhistorical phenomenon grounded on the "natural laws." Unable to explain the origin of rent, Ricardo developed a special case of differential rent through the arbitrary assumption of moving from better to worse land. Methodologically, both from the standpoint of Marxian theory and from the stance of reality, this type of differential rent is devoid of social and historical significance, and bereft of usefulness in decisions surrounding the exploration of oil today (see Bina 2013, Weeks 2013).

Thus, if rent is considered a *phenomenal* manifestation of the interaction of the structure of accumulation and the structure of landed property, it must be (1) a social category and (2) a category that is historically unique and specific to the capitalist mode of production. This Marxian criticism can be extended to the modern literature on "exhaustible" and "renewable" resources. From the standpoint of capital-in-general such a division is devoid of meaning and thus entirely arbitrary. This approach, in addition to having a firm foundation in commodity fetishism, neglects the cause of rent; the analysis is *ahistorical.* Under the capitalist mode of production, capital is self-expanding value. It expands into every facet of life, including the realm of nature for further accumulation. An element of nature, such as an oil pool, that did not have an exchange value based upon the generalized system of exchange in previous historical epochs, has now been transformed into an instrument of the expansion of capital. Therefore the significance of nature and, for that matter, *natural resources*, is not merely in their use-value form. Nature's importance is due to its utilization for the production of surplus value, thus its manifold forms are rather incidental.

## Marx's Theory of Rent

As we have pointed out, the notion of rent from Ricardo's viewpoint relates to the existing differential qualities of land in agricultural production or, for that matter, in mining. According to Ricardo, *the order of utilization* starts from the best land and gradually extends to the worst land as the demand for the product of land increases. This is also true for Sraffa and the Sraffian notion of *scarcity rent* (Sraffa 1960, Ch. 11). In passing, the Sraffian system, despite the façade of elaborate mathematics, does not run much beyond the preclassical Physiocratic theory in which rent stands for the attribute of nature—it is not a modern category that owes its social origin to the landed property. Returning to Ricardo, since the price of corn is determined by the conditions of production associated with the worst land that pays zero rent, the production from more productive land will result in a surplus over and above normal profit. That is the reason why Ricardo argues that "corn is not high because a rent is paid, but a rent is paid because corn is high." (Ricardo 1976: 38).

Accordingly, as the demand for agricultural products increases and the quality of land under cultivation deteriorates, the price of corn will be necessarily increased. Such an increase will also be accompanied by a higher and higher amount of rent over time.

Even though Ricardo correctly recognized the notion of differential rent in agriculture at the level of distribution, he essentially missed its significance in the production process and its relation to accumulation by accepting the Malthusian theory of population and its implication for the inevitable scarcity of agricultural products and food. In fact, the key to the conceptual presumption of the simultaneous increase in prices and rents is the assumption of relative scarcity in conjunction with the extension of cultivation from better to worse lands. Contrary to Ricardo's presumption, however, the development of the productive forces in agriculture is such that neither the best nor the worst land will always remain the same. An a priori assumption of the extension of cultivation from better to worse lands is not only conceptually false, but also historically without any empirical support.

In contrast to Ricardo, Marx approaches the phenomenon of rent by rejecting Ricardo's arbitrary assumption. The precondition for such a rejection was the concrete study of the *actual* direction of movement in the price of corn and the rent of land under cultivation during the first part of the nineteenth century in Britain. Pertaining to the above context, in his letter of January 7, 1851, Marx writes to Engels:

> I am writing to you today in order to lay a little question of theory before you...
>
> 1. There is no doubt that as civilization progresses poorer and poorer kinds of land are brought under cultivation. But there is also no doubt that, as a result of the progress of science and industry, these poorer types of land are relatively good in comparison with the former good types.
> 2. Since 1815 the price of corn dropped...Rent [nevertheless] has continually risen.
> 3. In every country we find, as Petty [Sir William] already noticed, that when the price of corn drops the total rental of the country rises (Marx and Engels 1975: 47–48).

The recognition of the above historical facts in conjunction with adequate knowledge about the work of his predecessors would seem to be among the factors that contributed to Marx's understanding of the subject and his distinction between (1) the *equal* application of capital upon land of *unequal* quality and (2) the *unequal* application of capital upon land of *equal* quality.

In addition to these findings in vol. 3 of *Capital*, the above distinction can be clearly seen in *Theories of Surplus Value* (vol. 2):

> *With a given capital investment,* the variation in the amount of rent is only to be explained by the varying fertility of the land. The variation in the amount of rent, *given equal fertility,* can only be explained by the varying amount of capital invested. (Marx 1968: 42–43, emphasis in original)

The above conditions are referred to by Marx as differential rent of the first type (DR I) and differential rent of the second type (DR II). It should be apparent that in the actual situation these two types of differential rents are formed jointly, in the presence of each other. In other words, the formation of social value is the result of the *unequal* applications of capital to *unequal* qualities of land through competition. Moreover, DR I and DR II are not in actuality formed independently of each other. The production conditions that lead to changes in the level of one form of differential rent do not necessarily leave the other form unchanged. Ben Fine summarizes the above problem in the following manner:

> For DR I, there is the problem of determining the worst land in the presence of unequal application capital (DR II)...for DR II, there is the problem of determining the normal level of investment in the presence of differing lands (DR I)...These problems concern the simultaneous determination of worst land and normal capital in agriculture. (Fine 1979)

To solve this problem, Marx analytically examines the formation of DR II in conjunction with DR I under changing prices of production. Moreover, instead of attempting to identify the simultaneous determination of *worst land* and *normal capital* directly, Marx let the prices of production decline, leaves them constant, and let them increase to capture the varying impact of capital investments. In other words, Marx successfully goes beyond the static distributional question of rent to address the question of the dynamics of capital accumulation in conjunction with the barrier of landed property in agriculture. At the same time, unlike Ricardo's, Marx's theory of differential rent is quite consistent with his theory of value, which is based on socially necessary labor time, as opposed to the labor-embodied varieties.

Another aspect of Marx's critique of Ricardo's rent theory is the rejection of Ricardo's notion of differential rent based upon the margin of cultivation along the "descending line." Marx has shown that, depending upon the conditions of production, there can be movements toward the more productive or less productive lands. After giving extensive examples, he remarked that, "In reality, the ascending and descending lines will cut across one another, the additional demand will sometimes be supplied by going over to more, sometimes to less fertile type of land, mine or natural agent" (Marx 1968: 273). On the contrary, Ricardo believed that the market value of agricultural commodities is *always* determined by the labor required on the no rent marginal land. The difference between Marx and Ricardo is great.

Marx argues that,

> As soon as the additional supply surpasses the capacity of the market, as determined by the old market-value, each class naturally seeks to force the whole of its product *onto* the market to the exclusion of the product of the other classes. This can only be brought about through a fall in price, and moreover a fall to the level where the market can absorb *all* products. If this reduction in price is so great that the classes I, II, etc., have to sell *below* their costs of production, they naturally have to withdraw [their capital from production]...But it is further clear that in these circumstances it is not the worst land, I and II, but the best, III and IV, which determines the market-value. (Marx 1968: 292–93, emphasis in original)

The above statement is clearly different from that of Ricardo. The competitive struggle of capital invested in different "classes" of land and the existence of specific market conditions can certainly lead to the formation of a market value different from the value of the product of marginal land. As a result, such a market value regulates the market by definition and is *below* the value of the marginal producer.

To illustrate the mechanism of differential rent, Marx assumes that the "marginal land" has a zero rent. However, he insists that such an assumption of zero rent is not essentially necessary (Marx 1981, Ch. 45). Furthermore, contrary to Ricardo, Marx notes that under the capitalist mode of production the formation of rent results from the development of landed property in agriculture. Ricardo, on the other hand, does not attempt to analyze the phenomenon of rent in association with the internal development of capitalist social relations. Instead, he tries to explain that rent is the result of differential fertility and other physical characteristics of land. Hence, the context of rent from Ricardo's viewpoint is *natural* and technical rather than *social* and historical. Consequently, it is hardly surprising that Ricardo's theory of rent becomes *universal* in character, projecting the notion of scarcity regardless of the institutional characteristics of production (Ricardo 1976: 39). As we have seen, according to Marx, the precondition for the formation of differential rent is the existence of differential productivity and the resultant differential profitability through competition. However, it is due to the particularities of landed property that the differential profitability turns out to become *permanent* in agriculture.

Before proceeding to the next section, we wish to clarify a subtle issue on valorization and *mutuality* (i.e., organic unity) and *distinction* of capital and landed property in production and through the mediation of nature. Contrary to Bryan (1990), the coexistence of "natural" conditions of land and conditions resulting from successive application of capital on land does not engender "incommensurability" between existing natural productivities and productivity according to the consecutive application of capital on land. This is so for two reasons: (1) that the landed property in its modern connotation is the creation of capital in the process of valorization and (2) that the raw "nature" (in this case, land) is a specific medium through which such a valorization is taking hold. Bryan (1990) dichotomizes these two contexts into "capitalist" and "pre-capitalist,"—that is, *social* and *natural*. He then proposes for nature, which is taken over by capital and eventually overrun by the successive employment of capital, and rent becomes a moot point. Bryan (1990) claims that once amortization of capital via the rate of interest is complete, there should not be any basis to speak of rent and rate of profit. Consequently, he chastises Marx for being preoccupied with the problem of transition to capitalism and not capitalism proper in the case of rent. He further suggests that we should euthanize rent exactly as neoclassical economics did a century ago. Bryan (1990) is mistaken for several reasons. First, the end result, the equilibrium-seeking attitude exhibited in Bryan (1990), leaves no room for the demonstration of contradictory dynamics in concept as well as corporeality of capitalism. He appears obsessed with neat reductions without limit. Therefore, he has little appreciation for a value theory (and rent theory) that spells out all these overriding contradictions at a more concrete level of theory and in the medium of the most tangible aspect of our life—nature. Second, he appears to be striving for an absolute theory of capitalism that will hoodwink the casual observer and remove any traces of fetishism, including commodity fetishism and, in the case of rent, fetishism of nature.

That is why any theory that is concrete enough to display the divisions (and further contradictions) within the capitalist subclasses is not his ideal. The phrase "landed property" troubles Bryan (1990) and smacks archaic in his mind, yet he does not bother to look around for its modern manifestation concerning capitalist class divisions and contradictions through the *mediating* presence of nature that does not vanish into thin air. On the one hand he removes landed property as the source of rent and, on the other, asks for "commensurability" between the "natural" and "artificial" sources of productivity. Hence his concern is not the latter but the very presence of the former. Third, capitalization (and amortization through the rate of interest) for Bryan (1990) is an *ideal type* that knows no limits, including its own. Hence he is consistent in his suggested emulation of the neoclassical template—the euthanasia of rent. Bryan (1990) writes: "Neoclassical economic theory has divorced the concept of rent from land and attached it to general conception of monopoly. Marxist theory could well do the same" (p. 180). Finally, Bryan (1990) appears to have a penchant for a capitalism that is perfect, unadorned, stripped of *nature* in us and around us; a robotic ghost reduced to amortization and the rate of interest; a capitalism of the neoclassical textbooks; an unembellished capitalism—a pure capitalism that is free of all nuances both in *theory* and *corporeality*. Thus a nuanced capitalism smacks precapitalism for him. That is probably why Bryan (1990) castigates Marx for offering a theory that is so concretized at a complex level of abstraction in rent.

In a somewhat similar context, it is instructive to return to *The Poverty of Philosophy* (Marx, 1973b, chapter 2, section 4:141–44) and to ascertain how Marx so vehemently contested the gist of Pierre-Joseph Proudhon's *The Philosophy of Poverty* and why he so disparagingly characterized him as a romanticist. Marx begins in his *Foreword* with the following words:

> M. Proudhon has the misfortune of being peculiarly misunderstood in Europe. In France, he has the right to .be a bad economist, because he is reputed to be a good German philosopher. In Germany, has the right to be a bad philosopher, because he is reputed to be one of the ablest of French economists. Being both German and economist at

> the same time, we desire to protest against this double error. (Brussels, June 15, 1847)

A short passage from Proudhon's should suffice that Marx anticipated, some 140 years ahead of Bryan (1990), the extrication of rent from land by neoclassical economics and its relegation to a "general conception of monopoly." Echoed somewhat in Bryan (1990:180), Proudhon's proposed "improvement in the use of the land" is as follows:

> Rent is the interest paid on a capital which never perishes, namely—land. And as the capital is capable of no increase in matter, but only of an indefinite improvement in its use, it comes about that while the interest or profit on a loan (*mutuum*) tends to diminish continually through abundance of capital, rent tends to always to increase through the perfecting of industry, from which results the improvement in the use of the land... Such, in its essence, is rent. (cited in Marx 1973b: 142)

The above passage has little appreciation for competition and the mutuality as well as the distinction of landed property and capital in valorization.

## Rent in the Neoclassical Economic Parlance

It is instructive to find out that in dealing with "practical problems," modern orthodoxy that had so readily detached rent theory from its core, prefers from time to time to employ some sort of scarcity rent (often as "user costs")[3] when it tends to confront the situations such as the ones that have been developed in the oil industry. But the concept of rent is a tricky one for those who have accepted the neoclassical general equilibrium. On the one hand, using the framework of general equilibrium (a simultaneous determination of factor incomes) one has to engage in the generalization of "rental income" for all the factors involved at the margin of production (Clark 1891, Hobson 1891). Hence, methodologically, the same principle governs the distribution of the incomes of all the factors involved in production. These factors are labor, capital, land, and entrepreneurship with respective returns of wage, interest, rent, and profit. On the other hand, given the above approach, there is no possibility of treating particular factors, such as agricultural or urban land, oil fields, coal fields, etc., specifically unless the above framework is replaced by that of partial equilibrium (Fine 1982a).

Within the context of partial equilibrium, it is possible to establish a *causal* relation for one factor independently of other factors that can be specifically differentiated from the general principle of the formation of other factor incomes. For instance, taking the above approach, Marshall treated the notion of differential rent in association with lands of different quality in agriculture (Marshall 1893). In so doing his usage of the partial equilibrium framework was quite consistent with his treatment of rent as a price-determined factor income. This type of analysis, however, is not without problems of its own. Moreover, within this framework, one has to assume either a one commodity economy or, in the case of a multicommodity economy, constant prices for all other commodities; the following passage is instructive:

> In a one-good world, rent would be price determined according to the differential productivity of better over marginal (no rent) land in use and a particular role could be assigned to land in causing differential productivity and hence rent as in Ricardian Theory. (Fine 1982a: 344)

Far from the heated debates that occurred among the contending factions of the emerging neoclassical school at the turn of the nineteenth century (and the early part of the twentieth century), a modern neoclassical economist of today, who is trained to think in terms of general equilibrium theory (hence, simultaneous determination of factor incomes), does not even begin to question the significance of the above trade-off as it pertains to the question of rent (see Krueger 1974, Foster 1981, Ng 1983, Devarajan and Fisher 1982, Wilson 1979).

Confronting the real world, these "modern theorists" would soon recognize that there are many unexplained elements left out of their standard competitive general equilibrium models. But, faced with the question of why rent has to be treated specifically (i.e., on a different footing, in accordance with its institutional underpinning), they often prefer to treat the matter externally (Miller 1973, Brown 1974). Historically, given "the conceptual specificity of rent . . . the debate over the rent theory [among the neoclassical economists] was a debate between partial and general equilibrium and to that extent a dialogue of the deaf" (Fine, 1982a: 344). With the gradual dominance of the general equilibrium approach in neoclassical theory, a specific theory of rent became *superfluous,* thus removing the feasibility of any dialogue (see Wessel 1967).

## The Oil Rent and General Equilibrium

For those who tend to question the fundamental basis of neoclassical theory the subject of rent still remains troubling. But what is more troubling is the message of those who stand on the fence, ideally fantasizing as if they can reconcile the differences between neoclassical theory and its *classical* counterpart by means of methodological compromises that undoubtedly promote nothing other than theoretical confusion. One such example can be seen in the treatment of oil rents by J. M. Chevalier (1976).

Chevalier starts out by defining "the oil surplus as the difference between the market price of a ton of crude oil sold to consumers in form of finished products and the total average cost incurred in discovering, producing, transporting, refining and marketing this ton of crude" (Chevalier 1976: 281). At best, dealing with the notion of oil rents within the general equilibrium framework, he does not seem to realize that such a treatment tends to destroy the specificity of his rent theory at once. In addition, Chevalier maintains that due to the oligopolistic structure of oil production, and the lack of perfect mobility of capital in this industry, one has to distinguish between two types of oil rents; (1) differential rents and (2) monopoly rents (Chevalier 1976: 283–85). Of course differential rents so defined are generated through differences in production techniques and natural qualities, whereas monopoly rents are said to be the result of the differential profit rate of oil relative to other industries (Chevalier 1976: 285). He then devises four different categories for differential oil rents: (1) quality rent, (2) position rent, (3) mining rent, and (4) technological rent (Chevalier 1976: 286). Finally, when Chevalier comes to evaluate the mechanism of price determination, he admonishes Smith, Ricardo, and Marx (and several other economists) for not being able to recognize what he alone has *discovered* about the relationship of cost and price with respect to rent—in particular on the relationship between cost and price of oil. Here is what Chevalier had to say:

> None of these economists...paid any attention to the determining influence of the cost trend. The price of crude oils tends to be in line with the development cost of the most expensive deposit when costs are increasing, and with the development cost of the least expensive one when costs are decreasing. (Chevalier 1976: 298, ft. 44)

Chevalier's treatment of rent is not only a mishmash of methodological misunderstanding on the meaning of rent but also a consequence of a misleading amalgamation of Marshallian cost categories and the classical and Marxian concepts of value and prices of production.[4] First, by abstracting from the phenomenon of property relations in the production of oil, he scarcely realizes that within the framework of general equilibrium the causative determination of the least as well as the most expensive oil deposits cannot be differentiated from each other. For general equilibrium is a framework of simultaneous determination. Second, aside from the difficulties of a general equilibrium framework, it is not clear why the price of oil should be either in line with the cost of the least or the most productive deposits alone—given the assumption of ascending or descending cost trends respectively—and not somewhere between the two. Besides, basing the price of oil on the lowest cost deposits a priori, one cannot help but wonder about the status of higher-cost deposits and the existing differential oil rents—both empirically and theoretically. Hence a troubling ambiguity in the origin of differential oil rent at the point of production.

Here, the formulation of "quality rent," "position rent," and "mining rent" poses a formidable problem from the standpoint of identification of the origin of rent in the production process. The difference of "technical rents" from the above "rents" is also unclear. More importantly, however, Chevalier's supposed oil rents cannot possibly assume the status of a *social category* for they are collectively devoid of any social or property relations—and without any specificity. Hence, the choice here—other than going back to Ricardo—is between Marshall and modern general equilibrium theory. That is why Fine's narrative about the predicament of modern rent matches this occasion:

> The passage to extinction of rent theory in neoclassical economics has meant that it has lived in the underworld of the profession, like a guilty conscience that is at its strongest when crime is committed but which fades with the passage of time only to reemerge sporadically and feebly. (Fine 1982b: 99)

What is also disturbing is the contagious influence of the neoclassical theory of competition (and likewise its theory of monopoly) on the remaining contemporary schools of economic thought, especially the ones that are seemingly opposed to the mainstream economic orthodoxy (see Weeks 1981a, 2013; Fine 1982b; Shaikh 1982; Semmler 1982; Bina 1985, 2013; Moudud, Bina, and Mason 2013). As we have demonstrated, the general equilibrium approaches to the determination of "factor incomes" place all factors on the same footing, thus treating all factor incomes as rents. The difficulty of this method will be compounded further when allowing for the realization of these rents in conjunction with market structures other than "pure competition" (see Bina 1985, Ch. 6). As can be seen, Chevalier has failed to develop a *specific* theory of oil rent even within his own framework.

The next step is to show that one cannot develop a viable theory of rent for the oil industry independent of the possible impact of the ownership of oil reserves and the condition of oil leases on the accumulation of capital in the oil industry (see Bina 1985, Chs. 5 and 8). In this connection, we have chosen to deal with Fitch's treatment of oil rent (Fitch 1982). Even though Fitch correctly points out the shortcoming of neoclassical theory in dealing with such practical problems as the oil crisis of the early 1970s, he nevertheless fails to make a distinction between the nature of rent in classical political economy and its counterpart in Marx (Fitch 1982: 20).

We have seen that Ricardo developed his theory of differential rent based on the differences in productivity that existed between lands of marginal and intramarginal qualities. Moreover, he maintained that the price of corn is always determined by production on marginal land, or land of inferior quality. Hence, Ricardian rent is price-determined rather than price-determining (Ricardo 1976, Ch. 2; Bina 1985 Chas 3 and 5).

Unlike Marx, Ricardo rejects the notion of absolute rent and with it the impact of landed property on production in agriculture. Instead, his primary concern was the distribution of surplus among the social classes (for specific analysis see Fine 1979). Therefore, Ricardo's theory of rent, being formed at the margin of cultivation, is independent of the structure of landed property in agriculture. Moreover, Ricardo's rent theory is not consistent with his labor-embodied value theory. Striving for *specificity*, Ricardo's rent can be conceptualized either in a one-commodity world or in a multicommodity world with the prices of other goods remaining constant.

Although Fitch is critical of the prevailing view of his time, he travels on the Ricardian road to the Ricardian nondestination. The veneer of Ricardo's rent theory and its semblance to that of Marx's is rather deceptive. Fitch fails, alas, to differentiate between the two very different notions of rent—the classical and the Marxian. He maintains:

> By contrast, the Classical/Marxian theory accounts for the price of Persian Gulf oil without any recourse to such *dues ex machina*. The cost of production is properly understood as *unequal* for all producers

> and the market price is regulated by the producers operating on the basis of the least favorable conditions who are able to clear the market at a market price equal to their marginal price of production. So the result here is that surplus profits tend to originate more in primary commodities than in manufactured commodities because the range of cost differential is greater. (Fitch 1982: 20)

Clearly, the above passage departs from Marx's method of analysis and his treatment of rent in agriculture. As we have demonstrated, contrary to the *margin of cultivation thesis*, Marx argues that any one-sided movement from better to worse land is only a special case in agriculture (Marx 1981, Part VI). Even though Ricardo's treatment of rent is specific, it is valid only within a partial equilibrium framework. One has to remember that the concept of "the margin of cultivation" in Ricardian theory had been made more general by the emerging Marginalist school, for the calculation of factor incomes, before its eventual replacement by general equilibrium (Fine 1982a). Methodologically, given the lack of consideration of the effects of landed property on the pattern of capital investment in agriculture, the Ricardian theory is caught in a dilemma of its own making: on the one hand, it loses its *specificity* if it departs from partial equilibrium; on the other hand, it remains static, restrictive, and unrealistic if it does not. Theoretically, the above theory remains ahistorical and depends on axiomatic treatment, as it fails to account for the institution of landed property and its mutual relation with the pattern of capital investments (Fine 1979).

## THEORY OF OIL RENT AND OWNERSHIP OF THE OIL RESERVES

In this section, we intend to summarize our previous theoretical propositions to develop a theory of differential rent for the oil industry. The phenomenon of economic rent as a category distinguished from normal profits is neither original to Marx nor specific to classical political economy. However, what made Marx's theory of agricultural rent different from his predecessors' theories is "the *specificity* of the analysis itself, not the category" (Shaikh 1981). The notion of oil rent in the oil industry is none other than the phenomenal form of the specific property relation that is unique to the oil industry. Historically, the *separation* of the ownership of hydrocarbon deposits from the ownership of the oil fields resulted in the emergence of this barrier within the accumulation process in the production of oil. In countries or regions where the ownership of the surface soil legally includes the subsoil, the owners of particular oil leases, that is, capitalist producers, are confronted with the obstacle of ownership of the oil deposits. This relationship remains the same even if the state as a legal institutional form of landed property owns the oil lands; owing to the establishment of capitalist institutions (Bina 1985, Ch. 3). The separation of ownership is part of a historical process that is realized legally through the act of lease contracts, concessions, etc. At the same time, theoretically, capital investments made by the owners of the subsoil are also subsumed under the separation of ownership of the subsoil from ownership of land. The owner of the land comes to appropriate rent, while the capitalist investor tends to appropriate normal profit.

It is critical to note that the oil concessions in the Middle East and elsewhere in the world had been formed and framed to span nearly some 60 years of duration—a lengthy lease, nevertheless. These leases were signed into de facto contracts that for whatever reason renewed from time to time. In fact, there had been frequent modifications and adjustments in these concessions. For example, the revision of the original D'Arcy concession (1901) and adaption of a modified version in the 1933 concession between the Iranian government and Anglo-Iranian Oil Company (AIOC) is a case in point. Hence there are simply no "ifs" and/or "buts" in the issue of who owns oil reserves in the ground. Yet, Daniel Yergin believes that "the concessionaires" (i.e., leaseholders within the IPC) owned the reserves. We read:

> One of the legacies of Mossadegh, nationalization, gave the Shah comparative flexibility [concerning the 1957 partnership with Ente Nazionale Idrocarburi]. In the other oil-producing countries, the concessionaires—the foreign companies—still owned the reserves in the ground. (1991: 504)

This above statement is untrue, given that: (1) such *leases* never assumed to be equal to ownership of the oil-in-place even for the IPC itself; (2) the seeming "flexibility" of the Shah of Iran had little to do with the aborted nationalization of oil in Iran and more to do with pockets of potential oil reserves unconnected with the 1954 Iranian Oil Consortium (IOC); (3) the nationalization of oil by Mossadegh was annulled by the 1953 CIA coup d'état and de facto provisions of the IOC; and (4) Mossadegh's legacy hinges on the ownership of the sovereign, which was also acknowledged as a right of self-determination by the UN General Assembly (see Hyde 1956). The last point is not only warranted but also has legal precedent in international practice. That is why the present writer recommended the recent renationalization of oil in Argentina (see Bina, *Asia Times*, 2012b). Yet, Yergin keeps repeating the same nonsense:

> In all the member countries, with the exception of Iran, the oil reserves in the ground actually belonged by contract to the concessionaires, the companies, thus limiting the countries control. (1991: 523)

Yergin does not appreciate the difference between the *de facto* control of oil and *legal* ownership of oil reserves (oil-in-place) with respect to these concessions. He appears to take writing-off of oil royalties (as income tax) and writing-in of oil reserves (as an asset)—an ad hoc scheme by the IPC—on their face value.

A study that was completed in the early 1970s concluded that there is a major distortion in the exploration of oil that primarily "results from a widely divided ownership of land in the United States" (Miller 1973: 415). This situation stems from the fact that the oil fields are often larger than the corresponding US oil leases that belong to the firm that made the discovery. The result is that the full benefits will rarely go to the primary discoverer. To substantiate this point, Miller goes into a long discussion of the extent of *fragmentation* of oil leases through the examination of portions of profits received by the main discoverer of the field. As a first approximation, he uses the production share of the largest producer of a field as the proxy of the firm's profit share. From this empirical work, it was discovered that "the percentage of the benefits from a well-received by the discoverer declines with the size of the field" (Miller 1973: 416). Consequently the barrier of *fragmentation* of the pattern of land ownership tends to move the capitalist investors away from engagement in new and larger oil fields that often require the assembling of large tracts of land prior to exploration. The above study also demonstrates that the "Fields under 500 acres accounted for 60.73 per cent of the [oil] fields but for only 14.43 per cent of the total area. It is again clear that most oil must lie in fields [that extend beyond] more than one ownership" (Miller 1973: 417–18).

Another problem is the fragmentation of oil leases in connection with secondary and tertiary recovery methods, where the whole field needs to be put under the control of a single management, to eliminate waste and enhance the productivity of the extraction process. This is called unitization. It would seem obvious that having a number of leases in a particular oil field undoubtedly works against production according to a predetermined schedule (Miller 1973: 423). The above condition demonstrates why the firms either move toward intensive exploration in the same areas, or simply concentrate on investing in the *existing* oil fields for further recovery. Even in the case of government-owned lands, due to the existence of noncompetitive leases (and at times the practice of granting inadequately sized leases to individuals through a system of lottery) there is a great deal of speculative activity combined with a considerable fragmentation of ownership in the US oil fields (McDonald 1971).

Faced with these obstructions, capital investment is made on exploration activity within the aged US oil fields, or aimed at further development of oil from existing oil wells, or canalized toward foreign oil fields. The comparison of the oil-well abandonment rate in the United States, during the periods of 1965–71 and 1971–74, reveals that there has been a tremendous decline in the rate of the abandonment of commercially exhausted oil wells in the latter period, even though the average life span of oil wells declined, from 26 to 24 years, respectively (see Bina 1985: 83, Table 20). This shows that although the average life span of producing wells during the period leading up to the oil crisis (1973) was shorter than that of the crisis period, the oil wells were not abandoned as quickly as they used to be. This condition indicates, that in the United States oil was largely produced through the *successive* investments of capital upon the already-producing US oil fields. However, it was not until the early 1970s that the US oil industry experienced a substantial decline in productivity, in terms of the average oil recovery per well, as these investments were further intensified (see Bina 1985: 71, 75, Tables 13 and 15). This can also be shown from the increase in the amount of development capital expenditures (per barrel), that is, those investments that are made upon the older US oil fields, during the periods of 1966–70 and 1971–74: an increase of 7 percent as opposed to a 261 percent increase, for the period leading up to the crisis (see Bina 1985: 75, Table 15). Meanwhile, the investment in the realm of oil exploration shows an increase of about 8 percent during the period of 1971–74.

The intensification of capital investments within the existing oil fields is by and large the consequence of the impediment of the prevailing pattern of land and lease ownership in US oil production. In this context, the structure of landed property and the fragmentation of oil leases played an influential role in the direction of capital investments and the structure of accumulation in the US oil industry, long before the oil crisis of 1973–74, that set a new basis for the formation of market values, rents, and market prices at the global level. We have demonstrated elsewhere that, within the global context, the prices of all other sources of energy, including coal, natural gas, etc. are *regulated* by the *value* of oil produced from the aged US oil fields (Bina 1989a).

Given the above property relations, the formation of social value involves a process of intraindustry competition. Depending upon the extent of differential productivity, there will be value transfers from one individual capital (individual production unit) to another that would manifest themselves as differential rent. The internationalization of oil production is the manifestation of this process at the global level. During the early period in the oil industry in the Middle East and elsewhere, essentially outside the geographical boundaries of the United States, the production of oil was through the colonial contracts and in the parlance of political economy originated in rudimentary introduction of capitalism and thus *formal* subsumption of labor (Bina 1985, Ch. 3). However, as the material foundation of capitalism in these social formations and also within the international oil industry has further developed, the production of oil had gradually assumed the characteristic of the *real* subsumption of labor under capital at the global level. In consequence, a social *value* at the global level has emerged that systematically intertwined with the formation of differential oil rents through global competition. Based on the analysis of the previous sections, the same distinction should be made between the Ricardian (and neo-Ricardian) notion of value for the so-called marginal oil producer, and the market value grounded upon the *regulating capital* in the Marxian framework. It is not always the case that market value coincides with marginal producer in the customary norm of today. Therefore, from the theoretical standpoint, it may not be difficult to appreciate why general application of Ricardo's specific "margin," prevalent in today's mainstream theory, is not a fruitful method in the analysis of petroleum or possibly any other industries.

At the empirical level, we were able to identify the US oil fields (lower 48 states) as the least productive region of the world (Bina 1985, Chs. 7 and 8). The US oil fields are also the most explored oil region of the globe. Since the effect of differential oil rent of DR I cannot be separated from differential rent of the DR II, determination of the least productive oil region, in connection with the identification of the normal size of capital investment, is an impossible statistical task. We have encountered the very same problem that Marx faced a little over a century ago in agriculture. Following Marx's method, we were able to identify the specific oil fields upon which the social value of the oil was formed (Bina 1985, Ch. 7). Due to the integration of oil production at the global level and the fact that US oil comes from the least productive oil fields, the social value associated with the above oil fields has become the social value of the entire international oil industry. Thus, it is through the decline in the productivity of the old US oil fields located in the lower 48 states (in the 1970s), that the magnitude of the newly formed social value has been restructured simultaneously with the reorganization of the entire oil industry through the *crisis*.

## OPEC Revenues and the Oil Rent

Having dealt with the historical development of the Middle Eastern oil elsewhere (Bina 1985, Ch. 3), we need to demonstrate here that—unlike the "rule of capture" observed in the US oil industry, which signifies the existence of readily established institution of private property together with an advanced form of capitalism—the early oil royalties and concessions in the Middle East and elsewhere were indicative of precapitalist, state ownership. This was associated with social relations that had to be transformed completely before the integration of those regions into the international economy became possible. In this connection, for instance, the fourfold increase in the "posted price" of oil during the 1973 crisis cannot be systematically and fundamentally explained unless there is an understanding about the following three interrelated historical developments that together gave petroleum production a distinctive character.

First is internationalization of the petroleum industry and *unification* of all the existing oil-producing regions of the world since the early 1970s. Second is the recognition of the characteristic of specific property relations (such as mineral rights and lease ownership) that are associated with the production of oil (regardless of their form, whether based on the "rule of capture" or on state ownership), and that are conceptually functioning as the foundation of oil royalties and rents. Finally, one has to recognize the effect of the intensification of capital investments within the least productive oil regions (such as the US oil region) that together with the above conditions set the stage for determination of higher production price for the entire oil industry from the early 1970s onward. Thus, contrary to the prevailing opinion, a fourfold increase in the oil "posted price" (a variable that is not the same as market or spot price) has been the reflection of all of the above determinations that objectively unified the industry through increased competition among the existing oil regions, and resulted in market prices that are tending to the high production cost of declining US oil fields rather than to the seemingly arbitrary decisions of OPEC. In other words, OPEC did what it did because the entire oil industry was at the threshold of a social transformation that practically revolutionized its institutional structure, and not the other way around. In fact, the unprecedented gesture of OPEC in the early 1970s (and beyond) is itself explicable through this transformation (Bina 1985, Ch. 9, 1990).

As indicated in the previous chapter, we divided the entire history of the Middle Eastern oil and other early oil-producing regions, such as Venezuela, Mexico, and Indonesia, into three episodes of (1) the early oil concessions (1901–50), (2) the era of transition and transformation of the 1950–72 period, and (3) the era of internationalization of production that completes the integration of oil-producing regions into the global oil structure and necessitates the formation of market values, differential oil rents, and market prices within the global industry. Historically, at the beginning of the twentieth century precapitalist social relations were still dominant in the Middle East, Latin America, and South East Asia. But the penetration of international capital, especially oil capital, into these particular regions was gradually gaining momentum. The outcome was the establishment of a system of oil concessions that laid the cornerstone of the oil industry in the Middle East and elsewhere. These concessions were made with more or less uniform principles for surrender of the oil property rights of the local authorities or of the states to a handful of powerful transnational oil companies or individuals from advanced capitalist countries interested in oil exploration in such regions (see Cattan 1967b; Bina 1985, Ch. 3).

A quick glance at the Middle Eastern or Venezuelan concessionary agreements of this early period reveals that, without exception, all of these contracts were unusually long in duration and related to large areas often equal to the size of the countries involved. This system of oil leases, apart from the form of ownership, was qualitatively different from the lease contracts in the United States, which were usually concluded for a fairly short time and over a much smaller area, and with relatively larger royalties granted to landowners. The terms of the concessions, such as the size and determination of oil *royalties* and additional payment to the contracting governments and ruling authorities, were also different from what we experience today. Royalty has been defined as a portion of oil extracted from the land, which goes to the individual owner of subsoil or the contracting government. This portion in majority of cases is determined at 12.5 percent.

At the outset, in practice, this royalty was calculated on the basis of a fixed amount of money per ton of oil (i.e., shilling, cent, per ton). Thus, the fixity of payment and its lack of connection with the market price of oil are among the distinguishing features of the concessions of this period. It should be realized that even though there were "profit sharing" clauses in some concessionary agreements, they have never been honored in practice by the companies (Cattan 1967b, Ford 1954, Mikdashi 1972, Rouhani 1971, Bina 1985). The characteristics of the above concessions stem primarily from the social dominance of precapitalist relations in the early years that in turn necessitated a *rudimentary* (undeveloped) form of oil rent or royalty (Bina 1985, Ch. 3). In sum, the early development of oil in the Middle East and elsewhere saw a *direct* political domination by the international oil companies, in conjunction with moderate financial terms, a static (fixed) mechanism of royalty payment, and the lack of any relationship between the magnitude of oil prices and the amount of royalties.

As international capital tended to further penetrate into the Middle Eastern oil and other regional oil structures, the corresponding institutions and social relations of modern civil society in this and other regions gradually began to develop accordingly. At the same time, the economic and political conditions that were conducive to the development of modern industry in general (since the Postwar period), and to the growth of the Middle Eastern oil industry in particular (especially since the Marshall Plan was to benefit from cheap Middle Eastern oil), led to sharpened contradictions and the strengthening of capitalist social relations within the entire region. Accordingly, as the production of oil increased substantially, and the center of gravity of proven oil reserves gradually shifted to this region, given the primary insistence of Venezuela, the terms under which oil property rights were granted had to be revised. This era (1950–72), that also saw the formation of OPEC, started with the abolition of the oil concessionary agreements and the establishment and spread of the regime of 50–50 profit sharing. But to be able to implement this newly devised system and determine profits, without exposing the actual profit pictures, the international oil companies did use the "posted price" as allocation mechanism—an as-if price for valuation of crude oil and its internal transfer within the cartel's international networks and subsidiaries. In this manner, the "posted price" served as a *variable basis* (responsive to the captive markets that had to be kept in line with production) for determination and calculation of oil revenues and other associated payments paid to the oil-exporting countries of the Middle East and elsewhere by the companies (Cattan 1967b, Mikdashi 1966, Rouhani 1971, Bina 1985).

By looking at the history of this period, that is, the period of 1950–72, one can recognize an increasing tendency toward market orientation within the atmosphere of increasing conflict between national and international capital. During this period, the undeveloped form of oil rent of the previous era in the Middle East, Venezuela, South East Asia, and elsewhere was gradually transformed into a much more developed concept of oil rent, responsive to the changes in the market values, spot prices, and the emerging market conditions that were compatible with contemporary capitalism. During this transitional period, the social relations of capitalism were perfected in the Middle Eastern oil industry and the oil industry in Venezuela and elsewhere. The extent of the socialization of production at the end of this period can be observed from the tendency toward global price formation and increased *competition* among the existing petroleum industries at the international level. This transformation was manifested in the formation of market prices based on the least productive region and in the formation of differential oil rents according to the existing differential productivity of the competing oil regions (Bina 1985, Chs. 6 and 9).

Since 1970, the decline in the productivity of US oil production on the one hand, and progressive integration of oil production within the global economy on the other hand, resulted in a higher magnitude of value, an increased volume of differential oil rent, and higher market prices globally. Given an increased level of differential productivity and profitability within the Middle Eastern, North African, South East Asian, and Venezuelan oil regions that naturally translates into an increased amount of differential oil rents, it is not hard to understand why OPEC demanded a fourfold increase in the "posted price" of crude oil during the 1973 crisis (Alnasrawi 1985, Bina 1985). The above analysis would also simultaneously account for the crude oil quality differentials and the transportation costs involved in global competition (Rifai 1974, Bina 1985).

## CONCLUDING REMARKS

We have shown that the notion of rent in political economy has been the subject of intense debates within the classical, Marxian, and neoclassical schools. It was pointed out that the Ricardian conception of rent has its origin in the technical conditions of production in agriculture (or other products of land). Even though Ricardo was very much concerned about the distribution of *surplus,* as Marx has eloquently shown, he was less successful in treatment of the impact of landed property upon the production process. Unlike Ricardo's, Marx's concept of rent is historically specific, and strives to understand the role of landed property in agriculture and its barrier to accumulation. He successfully demonstrates that capital accumulation in the presence of landed property is faced with obstacles that may or may not be easily overcome in the capitalist mode of production. Hence, to avoid the panoramic subterfuge, Marx needed to engage in a concrete investigation. Marx also holds that modern rent is very different from its medieval counterpart. Therefore, the significance of Marx's theory of rent is in its specificity.

With the advent of the "Marginalist School," Ricardo's rent theory reached its logical conclusion, through the invention of such concepts as *quasi-rent.* This was due to Marshall whose method of *partial equilibrium,* at least methodologically, has left the door open for a *specific* interpretation of the conception of rent in political economy. Meanwhile, we have seen that even though, from the standpoint of methodology, Marshallian theory underpins the neoclassical school, it nevertheless differs significantly from present-day general equilibrium analysis. One of Ricardo's major contributions is the distinction of rent from the category of profit. Marshall, however, blurred this distinction and thereby paved the way for his successors to eliminate it altogether from their theoretical models. This task was accomplished through a long and arduous struggle that led to a gradual acceptance of the general equilibrium approach to the problem of distribution, by putting all elements of production on equal footing, which automatically extended the rubric of *rent* to all the categories of "factor inputs" for good.

If the role of economic theory is to explain the nature and characteristic of economic and social institutions, one can only hope to grasp them in their complexity by uncovering their innermost fundamental relation rather than through the superimposition of ideal-types. Likewise, the treatment of rent independently of the effects of landed property misses the very essence of rent in capitalism by misplacing its cause. This is especially critical in understanding the literature associated with energy and oil economics, since the majority of the authors in the modern literature use the attributes of scarcity or *exhaustibility* as the primary source of rent. To be sure, since scarcity (and, by implication, exhaustibility), in the broadest sense, pertains to all historical epochs—including the hunting and gathering societies of the past—one can neither build a generalized price system nor a mechanism for allocation of rent specific to the capitalist mode of production. Thus, focusing on "exhaustibility" in the context of rent (particularly, oil rent), is no more than succumbing to fallacy, and hence circularity of fetish mongering within orthodox as well as heterodox economics of today.

Similarly, we argued that the claim of having a specific theory of oil or energy rent by these theorists is highly exaggerated. The majority of theorists have only *superficially* recognized the implication of oil rents, since their general equilibrium framework does not leave any room for a specific theoretical treatment of oil rents. As for others who have treated the notion of oil rent within the classical tradition, they failed to recognize the significance of oil rent from the standpoint of production, as they focused on distribution alone. Besides, for many writers oil rents emerge through monopoly and imperfection, whereas our concept of oil rent (especially after the globalization of oil) is consistent with increased capitalist competition. We have developed the theoretical basis of oil rent that has its roots in specific property relations that manifest themselves, in one form or the other, within the global oil industry.

# 3

# OPEC: Beyond Political Battering and Economic Romanticism

*It is impossible for someone to lie unless he [she] thinks he [she] knows the truth. Producing bullshit requires no such conviction.*

—Harry G. Frankfurt
*On Bullshit* (2005)

## Introduction

This chapter demonstrates that, far from being a cartel (or monopoly), Organization of Petroleum Exporting Countries (OPEC) is the mutation of an intimidated, insignificant, and idiosyncratic entity—in the semicolonial/semicartelized atmosphere of 1960—that turned into the organic body of global oil, following the collapse of colonial oil order in the embodiment of the International Petroleum Cartel (IPC) in the post-1974 period. The outgrowth of OPEC is the tale of two different settings. The former was semicolonial and controlled; the latter is postcolonial and competitive. These two settings are like night and day, with no chance of commensurability, either in theory or in practice, by *any* stretch of imagination. OPEC of the 1960s was a thorn in the side of the IPC, not so much for the fact that it may have shown a bit of resistance—not at all. This was due to the fact that OPEC was a paradox, a manifestation of an in-house diagnostic tool, that gently measured the magnitude of cracks and splinters that were already multiplying in the foundation of the Achnacarry. As we have demonstrated throughout this book, these cracks and splinters were the result of the whole host of factors and conditions that brought the Achnacarry down. OPEC was the face of oil royalty and merely a *symptom* of some of these deep-seated changes and challenges—such as rise of the independent oil companies that were already at the gate. Yet the *significance* of OPEC, not as an ironic face of this period, but as an organic part of the dynamics that put an end to the Achnacarry, should not be underestimated. The significance was displayed rather expectedly in the brief period of the crisis in the early 1970s. Since the 1973–74 oil crisis, we contend that OPEC's importance originates from the significance of oil rents, and not the reverse. Finally, our allusion to the tale of two settings may have yet another critical meaning that amounts to a preamble to further changes in the global polity. In the semicolonial and controlled setting of oil, the umbilical cord of US foreign policy was tightly connected to the livelihood of the cartelized oil and the IPC; in the postcolonial and competitive setting of the post-IPC, the cord was cut and US foreign policy has since been on its own. And it would be silly to think that anyone by any means would be able to switch these settings like the channels on a television.

As demonstrated earlier in this book, if differential oil rents are considered to be price-determined and if OPEC is an organization comprising the collectors of these rents, then the question must not be whether OPEC is a cartel (or monopoly); the question must be as to what is wrong with the adherence to a theory that has been falsified some tens of thousands of times, and its results are either unequivocally contradicting or equivocally inconclusive as to the alleged monopoly of OPEC. This is the real story behind the scores of failures by mainstream economic models with respect to the nature of OPEC.[1] Yet, alas, it does not occur to these economists that the reason behind these letdowns could be that—just like the proof of the existence of God—this particular "theory" is not in the least falsifiable. The ironic reality of OPEC, however, could be an arbiter of the first order to demonstrate the frivolity of the neoclassical theory of competition and dogmatic attitude of the many in orthodox economics tradition with respect to authenticity of competition in modern capitalism.

OPEC was established on September 14, 1960, in Baghdad. Here, in this meeting—and perhaps in the entire history of the organization—two individuals should stand head and shoulders above others in the leadership. These are J. P. Perez Alfonso of Venezuela and Abdullah Al-Tariki (al-Tārighi) of Saudi Arabia, genuine anticolonial nationalists at the time. The founding members of OPEC were Iran, Iraq, Kuwait, Saudi Arabia, and Venezuela whose share of oil royalties were directly tied to posted prices. The magnitude and variation of these "posted prices" were exclusively determined by the IPC. In September 1928, IPC was forged under the auspices of the "Big Three" (Standard Oil, Royal Dutch-Shell, and Anglo-Persian Oil Company), at the Achnacarry Castle in Scotland, to control the flow of oil from the gusher to the gas station. This arrangement is known as the "As Is Agreement" according to which nearly all oil across the international divide—from the stages of exploration, development, and production through to the international distribution, refining, and retailing—will have to be controlled by a handful of combined and amalgamated oil companies.

In a broad context, the founding of OPEC was a response to the unremitting cuts in the magnitude of "posted prices" motivated by a combination of factors, such as the 1958 economic recession, the expansion of Russian oil, and the imposition of the 1959 US oil import quota on the movement of oil from the Persian Gulf, which was seeping through rather progressively into the east-coast markets in the United States. The last factor was designed (1) to discourage competition from the independent oil producers through the restriction in the inflow of supply into the US market and (2) to limit the outflow of oil from the Persian Gulf region, thus complying with the tenets of the IPC's agreement at the Achnacarry Castle. Hence the trace of US government in cahoots with the IPC and the deliberate concealment by the former under the convenient cloak of "national security," was at the expense of US domestic consumers who paid a higher price at the pump and the Persian Gulf royalty-collecting nations that lost a chunk of their oil revenue both from the price cut and from the loss of production. At the same time, the production of oil in Venezuela also suffered from the US oil import quota in 1959.

Formation of OPEC was symptomatic of the initial fracture within IPC's internal structure, which in time was turned into widespread breakdown during the oil crisis of 1973–74. The crisis accomplished three things: (1) it decartelized the production of crude oil across the globe, (2) it globalized crude oil by means of competitive markets, and (3) it led to competitive appropriation of differential oil rent by the owner of oil reserves across the globe, irrespective of OPEC membership. Historically, posted prices functioned as an allocating mechanism for transfer and disbursement of crude oil within the IPC's worldwide networks. And since all markets were under the IPC's unmitigated control, given production costs and predetermined share of oil royalties, posted prices served as a benchmark for transfer of revenue from one pocket to another across the undifferentiated coffers of the IPC.

OPEC was not only *seen* as an economic challenge germinated within the IPC's international network, it was also *seen* as a political challenge against the IPC-friendly US foreign policy and the prevailing political order under *Pax Americana* (1945–79). Consequently, even minor disputes against the IPC's unilateral control by OPEC were automatically seen as a potential threat by the US and British governments, despite the very fact that the majority of OPEC members had either been allied with or as a client-state accepted the precept of US hegemony. Yet, astonishingly, neither the United States nor Britain has attempted to recognize OPEC officially before nearly a half a decade after its formation. Indeed, this consistency in the maintenance of the oil status quo is remarkably reflective of the 1953 CIA overthrow of Mossadegh, who nationalized Iran's oil but, at the same time, sought cordial relations with the United States.

By the end of the 1960s, competition from the "independent" oil companies signaled the futility of effective control over the exploration, development, and production of crude oil across the globe. This, irreversibly, undermined the worldwide control and administrative pricing of oil and eventually led to de facto dissolution of the IPC in 1972, prior to the oil crisis of 1973–74. OPEC's role in this dissolution, however, should not be seen just as its apparent outlook as an active agent, but as the symptom of evolutionary transformation of oil toward unified pricing and competitive formation of differential oil rents. OPEC not only objected to falling "posted prices" but, as time passed by, to their arbitrary determination by the IPC.

In October 1973, OPEC doubled the "posted price" of oil exported, by the member countries, from less than $3.00 to $5.12. Oil prices were beyond the $20 mark on the spot market in the period between October 1973 and January 1974. The posted price, for Light Arabian crude, was increased once again to $11.65 in January 1974. This unilateral act coincided with the 1973 Arab-Israeli War, which also prompted a brief symbolic embargo against the United States (and the Netherlands) by a handful of OPEC Arab oil producers, some of which were US client-states in the region. The embargo was a deliberate response to rearming of Israel during the course of the conflict by the United States. Yet, in retrospect, it was also the first warning-shot that hinted the end of an era and the beginning of a new one, which permanently put an end to US foreign policy in cahoots with the now defunct IPC (1928–72). Yet, the specter of early oil conspiracies (and cartelization) is still haunting many who tend to uncritically reproduce the past in the presence of their imagination.

"OPEC's mission is to coordinate and unify the petroleum policies of Member Counties and ensure the stabilization of oil markets in order to secure an efficient, economic and regular supply of petroleum to consumers, a steady income to producers and a fair return on capital to those investing in the petroleum industry" (OPEC

Website). However, since the crisis of the early 1970s that irreversibly led to decartelization and globalization of oil, OPEC's mission statement of "stabilization of oil markets" does not lend itself either to theoretical scrutiny or to empirical verification. Ironically, this baseless claim reinforces precisely the alleged monopolistic view held by OPEC adversaries, who have now turned the table and seized upon this rhetoric in the media, despite ample evidence to the contrary. The pity is that across the board there is little difference between the tutored and untutored opinion on this pertinent issue.

OPEC is neither a cartel nor a monopoly, aside from its own confusing mission statement and the allegation of its adversaries. The productions of oil from non-OPEC regions, namely, Norway, Mexico, the United States, or the United Kingdom, are also entitled to differential oil rents that are in accord with global oil spot (and futures) prices. As for the practice of "swing production" by Saudi Arabia, it was proved self-injurious to the Kingdom and without any remuneration to any producer worldwide, particularly the least productive US domestic oil producers. The swing production by Saudi Arabia, which was instigated by the Reagan administration to weaken Iran and bolster Saddam Hussein during the Iran-Iraq War, and which led to rock-bottom oil prices (some $9 per barrel), was neither technically sustainable by the ARAMCO nor affordable by the Saudi economy nor even economically acceptable to the US producers in the oil patch, who permanently capped a substantial number of their productive oil wells in the mid-1980s.

Given the extent of muscle-flexing essentially initiated by OPEC "radical" oil exporters in the early 1970s—namely, Algeria and Libya, and, to some extent, Iraq—the road to negotiation between OPEC members and the IPC was paved on an irreversible course. And seemingly submissive members—namely, Iran (under the Shah), Kuwait, and Saudi Arabia—benefited from spillovers of genuine protagonists. However, OPEC's real power was truly dependent upon the underlying forces that regulated the price of crude competitively in the budding global spot markets. Yet, OPEC members overplayed their hand on price determination, in part due to their inability to see that their power resides in differential oil rents (not OPEC as façade), and in part for OPEC's leading role in the transition from cartelization to competitive globalization. From the early 1980s onward, in the presence of full-fledged spot (and futures) markets and deeper forces of (competitive) globalization, OPEC's contradictory posture eventually backfired.

Today, OPEC's (spot) basket of 12 types of crude oil compatibly responds to deeper forces underlying the globalization of oil by taking its cue from global oil spot (and futures) markets—notably from Brent (London) and NYMEX (New York). OPEC members include Algeria, Angola, Ecuador, Iran, Iraq, Kuwait, Libya, Nigeria, Qatar, Saudi Arabia, United Arab Emirates, and Venezuela.

## The Review of Literature

From the theoretical standpoint and from the standpoint of truth, it is critical to recognize that the majority of the writings on OPEC are strung together by the thread of neoclassical competition. This thread is far and wide, and runs through the writings by specialists as well as populists and journalists. For instance, the title of a recent *New York Times* article on the June 2012 OPEC annual meeting runs: "Despite Price Drop, Oil Cartel Keeps Production Limit" (Kramer 2012).[2] Truth of course is the first casualty of this passive war for perpetuation of theoretical banality and modeling based upon an absurd notion about competition, and then reversing the course on the authenticity of capitalism and chastising OPEC (and globalized oil) for being a genuine part of the whole. In this critical sense, the literature that follows could not be more misleading.

Neither the lumpiness of capital nor the size of the oil deposits located underneath the countries that assembled in OPEC would change the fact that capitalist competition—far from the fantastic imaginary in mainstream economics—is none other than the contention of capital against capital in the process of capital accumulation. This contention is in essence an outcome of the concentration and centralization of capital—a very tangible matter and the lifeblood that keeps capitalism going. This contention also intensifies with the development of capitalism toward maturity. There is no capitalism without this contention, except as a figment of imagination. Thus the conundrum concocted by neoclassical theory and adopted by mainstream economics is a fictitious puzzle with fantastic solutions.[3] This conundrum is to set the contest in a dichotomy between "perfect competition," on the one hand, and "imperfect competition," oligopoly and monopoly, on the other. This is a false dichotomy that is potential for scorn by many methodical scholars in scientific disciplines. Capitalist competition is not a state of affairs to be perfect or imperfect; it is a dynamic process, and with further growth of capitalism it becomes ever more intense and disposed to concentration. Thus the lumpiness of capital is the sign of competition, not monopoly.[4] Likewise, while subject to valorization of capital, it would be preposterous to assume that oil deposits found in huge lumps beneath the geography of particular countries are a sign of monopoly or "imperfection" of competition, and thus monopoly of OPEC. In this context, the review of the following selected samples (and models) below may shine a beam of light on the vast and unquestionably discouraging mainstream economics literature on OPEC.

The models reviewed below can be roughly divided into two groups: cartel (including the dominant-firm models) and noncartel models. Cartel models consider OPEC a profit-maximizing organization, seeking monopoly profits through the manipulation of its output. These models try to demonstrate the dominance of OPEC within the oil market resulting from collusion among its members. Noncartel models, on the other hand, contend that the crude oil market is competitive (albeit within the neoclassical economics paradigm) and the variations in price of oil can be explained by factors other than cartelization. An early review of these models can be found in Fischer, Gately, and Kyle (1975) and Gately (1984, 1986).

Typical cartel models, for instance, by Blitzer, Meeraus, and Stoutjesdijk (1975), Pindyck (1978), Ezzati (1976), Adelman (1980, 1982)[5], Salant (1982), and Aperjis (1982), all emphasize the dominant role of OPEC in the oil market and attempt to describe OPEC as a monopoly, oligopoly, and thus a cartel. These models assume that OPEC members have a unified goal and thus collectively set the price of oil in the global market. Hnyilicza and Pindyck (1976) divide OPEC into two blocks, "spenders" and "savers," before seeking an optimal (game-theoretic) bargaining solution within the alleged cartel. Heal and Chichilnisky (1991) focus on intra-OPEC bargaining.[6] Thomas (1992) looks at the "cartel stability" and "punishment strategies" via a common-pool exhaustible resource game. All these models, nevertheless, try to predict the future direction of oil prices by watching OPEC closely.[7]

Cremer and Salehi-Isfahani (1980, 1989) and Teece (1982), among others, present a competitive view of the oil market. These authors argue that the world oil supply curve is backward bending due to the hypothesized limited absorptive capacities of oil-exporting countries within OPEC, particularly those in the Persian Gulf. It was alleged that these countries would produce enough oil to meet their target revenue as required by their internal disbursement. As a result, a large upsurge in the oil price may reduce their need for further production. Consequently, these authors argue, that high oil prices during the crises of 1973–74 and 1979–80 could just as well have resulted from competitive equilibria (again in the neoclassical parlance), which might have eventually occurred without OPEC. However, the assumption of the so-called limited absorptive capacity, which stands as an analogue for "petrodollars recycling," has been proven invalid by the historical conduct and mere action of these oil exporters within OPEC. As was the common knowledge in the 1970s and 1980s, these countries not only grabbed and spent the hefty share of their windfall in the form of oil rents but spend a considerable amount of money also on unproductive activity—including the purchase of weaponry incommensurate with their own defense but as a handsome subsidy toward the US policy in the region—that they had to borrow beyond their own earning "capacity." Therefore, on empirical grounds, pointing to "capacity constraint" is not a persuasive argument. On a theoretical level, however, the argument of "limited absorption capacity" boils down to an idealized notion of national economy in marginal transformation and free of uneven development, internal class competition, violent capital accumulation, and, above all, to a self-contained, idealized economy immune and isolated from the forces of globalization. Besides, undue emphasis on the limitation of "absorption capacity" would shift attention from the sphere of production and (global) price formation to the sphere of circulation and distribution with respect to the disbursement of the amassed oil rents. Hence this "competitive view" of OPEC, not unlike the redundancy of *absolute space* and Newton's Bucket in motion, cannot stand on its own (for this parallel see Greene 2004).

Mabro (1975), MacAvoy (1982), Mead (1979), and Johany (1980) are among those who argue against the cartel theory by invoking other reasons for oil price variations than cartelization. Mabro (1975) ascribes the price hike of the early 1970s to the "less than perfectly elastic demand curve" facing OPEC and "the interdependence of [oil] production and future availabilities." His first point is an echo of "imperfect competition" in oil, his second point amounts to circular reasoning.[8] MacAvoy (1982) attributes the increase in the price of oil to supply disruptions. In particular, he points out that the 1973–74 oil embargo against the United States (and the Netherlands) led to the reduction of the world oil supply, thus causing the oil price to rise. In the same manner, MacAvoy argues that the 1979 Iranian Revolution and the 1980–88 war between Iran and Iraq reduced the OPEC oil supply and thus naturally raised oil prices.

MacAvoy's (1982) reasoning could be extended to the period following the 1990 invasion of Kuwait by Saddam Hussein and the beginning of US invasion of Iraq in the early 1990s. Contrary to such reasoning, Adelman (1982) contends that Saudi Arabia alone had the capacity to compensate for the production losses in the late

1970s (and by extension, in the 1990s), yet, Adelman argues, Saudi Arabia did not use its capacity and this consequently lead to a price increase. This reasoning is fallacious, because in another similar instance, that is, the shortage resulting from the Iranian Revolution had been decidedly compensated by other non-OPEC suppliers. Yet, MacAvoy's explanation falls short of explicating the cause of the price crash of 1986.

Finally, Mead (1979) and Johany (1980) attribute the cause of higher oil prices to a differential time horizon, that is, the transfer of property rights from foreign oil companies (with high discount rates) to host governments (with low discount rates). Here, speaking of "discount rate" trivializes the issue surrounding the transformation from cartelized oil to competitively globalized oil. First, other than treading on the surface of this cliché, the IPC has been an amalgamated body that had forged and acted in concert with a network of constituent companies in colonial and semicolonial settings; and this unquestionably would not qualify the oil cartel to be put in the same class as that of the ordinary enterprises, which operate in a much shorter time horizon for recouping their investment. Therefore, the discount rate attributed to the IPC's time horizon, as opposed to the one that supposedly fit the behavior of the exporting countries, is unwarranted. It is remarkable that these authors do not appreciate the *duration* and terms of oil *concessions*, with regard to *time horizon*, which continued until the very collapse of the IPC in 1972. Furthermore, given a wealth of historical evidence, the IPC rather had been dutifully practicing what is now called "sitting on concession" in Iraq (by the Redline Agreement), Iran (through frequent adjustment of supply by the Anglo-Iranian Oil Company and its successor Iranian Oil Consortium), and elsewhere in solidarity with the very term and tenet of the Achnacarry. Indeed, aside from the cosmetic changes, such as the so-called 50–50 profit sharing, the length of oil concessions remained unaffected. Second, in this case, a blanket reference to "discount rate" obscures the distinction between *profit* and *rent*, and thus muddles the historic transition of oil from cartelization to decartelization and globalization of oil. Mead (1979) and Johany (1980) thus would leave us no choice other than to go back to the realm of neoclassical guesswork once again.

In addition to all this, the disputes over the question of exhaustibility, intertemporal exploitation, and the extent to which "scarcity rent" may play a part in the cost and price of oil created added points of contention in the mainstream theory (see Gray 1914, Hotelling 1931, *RES* Symposium 1974, Pakravan 1976, Dasgupta and Hill

1979, Adelman 1986, 1990). Despite the apparent differences, all these models have their origin in well-known dichotomy of competition and monopoly, a dichotomy that is so pathologically ingrained in the standard mainstream theory.

## THE EVOLUTION OF GLOBAL OIL

The mere discovery of oil and domination of a handful of capitalist enterprises over the majority of the global oil reserves would not by themselves make the production of oil subject to the regulation of a fully established capitalist world market (DeNovo 1956, Ise 1926, Krueger 1975). The fact that substantial oil reserves in the less-developed regions of the world, such as the Middle East, North Africa and Latin America, were discovered by the pioneers of capitalist internationalization, does not necessarily imply that all these economies should have automatically become subject to the regulating mechanism or law of motion of capitalism. In fact, it has taken some of these countries more than half a century of development toward capitalism, however unevenly, to be able to successfully emerge from their impeding precapitalist social relations. Indeed, there are many countries that are still engaged in such a transformation and, at the same time, are caught in the contradictions particular to the uneven structure of internationalization under the auspices of the IPC, which persistently continue to this date in the concentrated economic sectors of such developing economies (Engler 1961, Penrose 1968, Jacoby 1974, Sampson 1975).

Thus, without adequate periodization of the history of the international oil industry, including the recognition of the characteristic of each period and its significance in development of the succeeding periods, the present conditions in the oil market cannot be adequately understood (Bina 1985). Given the monumental volume of oil literature, especially since the early 1970s, there is hardly any agreement among theorists on the methodological treatment of the subject, apart from an axiomatic and a historical orientation, which excludes simple recognition of the institutional context. In other words, while there is a great deal of discussion about the history of oil or OPEC, when it comes to theorization and conceptualization, there is a marked dichotomy between concepts and their historical counterparts, between theory and its application (e. g., Fischer, Gately and Kyle 1975).

The early period in the development of oil, which coincided with the internationalization of finance capital, also coincided with the era of cartelization of the industry, both in the United States and abroad. The cartelization of US oil was an institutional imperative, so much so that, even with the 1911 antitrust landmark in the United States, it has maintained its significance for well over half a century (Davenport and Cooke, 1923, O'Conner 1955, Hidy and Muriel 1956, Keysen and Turner 1959).

The control of US domestic oil continued for many years through the calculation of the market demand factor under the direction of the Interstate Compact Commission, as a significant quantity of oil reserves was discovered in the East Texas oil fields (Blair 1976). Consequently, the oil companies were able to stabilize US wellhead oil prices, which in turn set the "posted prices" at the Gulf of Mexico. Given the fact that the wellhead at the Gulf of Mexico was the basing-point in pricing of all oil, control of US domestic oil and cartelization of international oil were the sine qua non of total control of the industry globally.

From the standpoint of leasing the subsurface for mineral exploration, there have been two significantly different institutional traditions in practice: (1) the rule of capture, the recognition of private ownership of land together with the ownership of mineral deposits in the subsurface, which has characterized the legal form of US oil and mineral ownership and (2) the state ownership of subsurface mineral deposits, the legal form of property ownership in all other oil-producing states. These two systems of reserves ownership have existed side by side in the oil industry up to the present time. In the case of the US ownership of public lands, the mineral resources located in the subsurface are accordingly owned by the same public sources.

The consequence of either of the above ownership structures, in conjunction with the natural differentiation of deposits, which is necessarily subject to capital accumulation in the oil industry, is the formation of differential oil rent, a phenomenon that results from the transformation of surplus profit through *intraindustry* competition. Thus, the more developed the extent of market relations (i.e., further commoditization through *interindustry* competition), the more established the formation of an institutional characteristic that leads to differential oil rents according to differential profitability in the entire industry (Bina 1985).

Globally, given the uneven development of capitalism, the international oil industry has come to embrace both advanced and less developed regions of the world simultaneously. This global network has extended from the most industrialized capitalist countries of the West, such as the United States, to the least-developed transient precapitalist societies of the Middle East and elsewhere. Thus, given the existence of the various socioeconomic formations within the global reach of such enterprises, one has to rely on adequate periodization to address the causes and circumstances of the institutional transformation of production and pricing of oil in the twentieth century. Hence, we need a historically specific explanation that will at the same time contain its own limits.

Toward the end of the transitional period of 1950–70, world oil saw a series of events that have transformed the relationship between the IPC and OPEC. From the 1970 Tehran conference to the two-stage quadrupling of "posted price" of oil during the 1973 embargo, all the rent-collecting states of OPEC were able to challenge the IPC to boost their share of oil rent. This challenge was preconditioned by the larger-than-life force of globalization that had swept across the world and the region, and at this time took a very concrete political form in the oil industry. Thus, the oil-exporting countries within OPEC have been able to increase their share of oil rent via the increased magnitude of posted prices; at the time of the crisis, however, these posted prices had been taking their cue from the budding and sporadic spot markets with genuine competitive prices at the global level (Minard 1980). In fact, in the early 1970s, the existence of the market prices in the embryonic form of spot oil markets was the sign of the imminent demise of the IPC and the birth of global competition in the oil market, owing to the transnational development of the industry.

This is indeed the beginning of a new era, an era of decartelization, which follows the mechanism of its own for the *law of one price* oscillating around value and centered on the *regulating capital* invested centered on the costliest oil region of the world by means of global competition. Given the formation of global prices, the producers in more productive oil regions, in addition to normal (competitive) profit, are in a position to capture their differential profits in terms of differential oil rents. In other words, the OPEC price increase in the 1970s was a response to the reorganization and rationalization of the industry as a whole, which necessitated a large increase in the magnitude of industry-wide production price oil across the globe. The result has been the formation of differential profit as a potential source of differential oil rent. The higher the production cost of the least productive oil region, the higher the magnitude of differential oil for the more productive oil regions.

As can be seen, OPEC has been able to collect the differential oil rents of its members merely based on the law of one price and according to the variation in the magnitude of differential productivity of its oil. Thus, OPEC revenues are not an arbitrary sum, as the orthodoxy in the economics profession alleges. These revenues mirror the magnitude of differential productivity and differential profitability of oil owned by OPEC members. And if tomorrow OPEC closes its doors for good and the members were to depart and go about their business, commensurate with the productivity of their oil, they would earn the same. Hence, contrary to the melded orthodoxy in the literature (see Stork 1975 and Tanzer 1974 on the left; Bobrow and Kurdle 1975, Griffin and Teece 1982, Salant 1982, and Gately 1984 on the right), OPEC is better understood, both at theoretical level and empirical levels, through the analysis of this book.

## Stages of Global Oil Development

As demonstrated in Bina (1985), the entire history of international oil during the twentieth century has gone through the three distinct stages of development. These are: the era of the early oil concessions, the era of "50–50 profit-sharing," and the era of unified global commoditization, and internationalization of the oil industry. On the domestic side (i.e., the US oil industry), one can recognize three different stages as well. The first stage in the development of oil in the United States saw a gradual emergence of horizontal and vertical integration in the industry that led to the formation of trusts and subsequent trust-busting tendencies, culminating in the antitrust law of 1911 (Keysen and Turner 1959). As we may observe from the history of oil production during almost three quarters of the twentieth century, both in the United States and elsewhere in the world, the cartel-like practices and the process of cartelization emerged as a formidable de facto institution (Stocking and Watkins 1948, Stocking 1950, De Chazeau and Kahn 1959, Blair 1976, Engler 1977). The second stage in the development of the US oil industry is the neo-cartelization era of production control, which culminated in the establishment of the Interstate Compact Commission. This period, which emerged subsequent to the trust-busting landmark of 1911, ended in the early 1970s, with the emergence of the 1973–74 oil crisis that transformed the entire industry irrevocably. Finally, there is the era of transnationalization of the oil industry (since 1974), which has rendered the US oil industry an inseparable part of the organic whole of global production and exchange in global competition.

## Colonial Concessions and the Cartelized Oil

In addition to the peculiar terms of the contract conducted between the rulers of the precapitalist oil rich nations and social entities in the Middle East and elsewhere—which, of course, put the question of national sovereignty on the front burner for a while, and later was counterbalanced with attempted oil nationalizations—in this early era, we can observe four additional historically specific characteristics (1) the existence of the concept of constant royalty in terms of a fixed sum per quantity of output, (2) the formation of joint ventures among oil companies for the purpose of joint control of new concessions, (3) the founding of the Achnacarry Agreement, and (4) the agreement of "posting" the price of oil, from anywhere in the world, based upon the wellhead price of oil at the US Gulf (of Mexico).

First, in contrast with the notion of economic rent that fundamentally emerges from the appropriation of surplus profits by virtue of the ownership of land or mineral deposits in the subsurface, the oil royalties of the early period were established on an arbitrary basis, often at four shillings per ton of crude oil, quite unrelated to its value or price (Cattan 1967a, 1967b). This is a significant phenomenon that should be considered as a first step, before the transformation of the oil industry in the Middle East, North Africa, Asia, and Latin America from a precapitalist to a growing capitalist industry, especially in the view of subsequent oil nationalizations in Mexico (1938), Iran (1951), and Iraq (1972).

Second, the early joint ventures among the transnational oil companies were not motivated primarily by concern over risk and capital requirements. Instead, these arrangements were the main instruments of joint control over substantial parts of the globe; they were intended for the direct control of the pace of utilization of oil reserves by a handful of private enterprises. The 1927 Redline Agreement by the IPC partners pertaining to the Iraqi oil concessions is a case in point (Blair 1976). It is noteworthy that the division of the world's oil resources has not been done solely according to individual company ownership, which implies the existence of a simple form of rivalry. The pattern of joint ownership has also been dominant, such as the collective agreements pertaining to the Saudis' Aramco or the Iranian Consortium (Cattan 1967a, 1967b). Thus, one cannot rule out either rivalry or cooperation in the process of cartelization during this period.

Third, another feature of this early period is the necessity of discretionary agreements. Even though it is difficult to report all instances of the worldwide cartel-like arrangements among the major oil companies, the fully documented "As Is Agreement" of 1928, better known as Achnacarry Act, reveals a great deal about the collective effort of the companies in this period (Federal Trade Commission, International Petroleum Cartel 1952).

Finally, for the purpose of pricing of direct intercompany oil transfers, the companies designated a set of administered prices called "posted prices." These administered prices were initially based on the production cost of crude oil in the Gulf of Mexico, formerly the center of gravity of international oil. Even though the share of Middle East oil production has increased significantly, as the center of gravity of the world's oil reserves had shifted to the Persian Gulf, the IPC had routinely charged the freight costs from the Gulf of Mexico even for the shipments that originated from the Persian Gulf (Adelman 1972, Blair 1976, Alnasrawi 1985).

However, as early as 1941, this practice was questioned by the British Admiralty. Similar objections were also raised, shortly after the Second World War, by the US administrators of Marshall Plan in Europe. Thus, the companies were forced to get rid of this phantom freight through recognition of the Persian Gulf as a second basing-point in addition to the US Gulf. But, as the cheap Middle Eastern crude was able to penetrate further into the Western markets, the companies soon recognized that, because of the significant difference in the production costs of these two basing points, the establishment of a single global posted price might not be possible. This led to the birth of a double basing-point system, with two different cost structures, which included both Gulfs. Accordingly, there were different posted prices for these oil regions. Nevertheless, the Persian Gulf posted price was more reflective of the US oil cost-structure than of its own cost of production. In other words, the cost of oil from the Persian Gulf, including the cost of "royalties," has been a fraction of the cost of its counterpart at the Gulf of Mexico. This comparison is also relevant to the production costs at the Venezuelan oil fields, which has been substantially below the US costs, but nevertheless kept according to the cartel's yardstick at the Gulf of Mexico (Mikdashi 1972, Issawi and Yeganeh 1962, Frank 1966, Leeman 1962, Rifai 1974, Tanzer 1969).

## The Regime of the "50–50 Profit-Sharing"

In the late 1940s and the early 1950s, there was a marked change in the political and economic conditions of most Third World countries. There were some fundamental transformations in the entire world economy, which set the stage for the beginning of a new era. In this postwar period, significant changes were also apparent in the international oil industry, and the nature and terms of oil contracts, which had already been the subject of a vigorous nationalization in Mexico (1938), and which were facing similar attempts in Iran (1951) and, much later, in Iraq (1968–72) (for Iran's case, see Elwell-Sutton 1955, Walden 1962). The significant characteristics of this period (1950–72) were the transition and transformation of the entire industry from the status of a cartel-like economic entity based on the administered pricing decisions, demand management, and the supply manipulation, to a world industry with an emerging unified competitive market, and prices that were subject to the regulation of a more developed form of capitalism on the world scale. As expected, in periods of transition such as this, one may find an assortment of fading elements of the old along with the embryonic but, nevertheless, developing constituents of the new. Some of the more distinct characteristics of this period were: the existence of long-term contracts, the establishment of the Persian Gulf basing-point system and subsequent designation of lower posted prices for its oil, the utilization of posted prices for the calculation of oil royalties and rent, and the formation of OPEC.

In this period, the control of oil was founded upon a double basing-point system, one associated with the cost of production of oil at the Gulf of Mexico and another based upon the cost of oil in the Persian Gulf (see Clark 1938, Smithies 1949, Keysen 1949, Stocking 1950, Machlup 1949). It is true that, initially, the IPC engaged in posting one "price" for both these oil-producing regions. But, as the Persian Gulf oil entered the markets in the Western Hemisphere and tended to displace US oil, the IPC was obliged to cut the posted prices, thereby admitting openly, but indirectly, an enormous margin of profitability in the IPC's Middle East operations. Paradoxically, however, cuts in the Persian Gulf basing-point posted prices led to an accelerated displacement of Western oil by its Middle Eastern counterpart. During this period, the long-term colonial contracts had remained intact. And there was neither a spot market, nor, for that matter, a futures market in crude oil (Prast and Lax 1983). Contrary to the previous period, the calculation of the oil royalties/rents in this period was based upon posted prices rather than a fixed sum of money per barrel of oil. Therefore, the magnitude of oil royalties/rents was calibrated according to the variability of posted prices, which is theoretically a tangible step toward a *price-determined* economic rent in the oil industry (see also Lowinger and Ram 1984, 695, ft. #8). As we shall see below, with further transformation of the industry since 1974, given the emerging global spot markets and globalization of the oil industry, *posted* prices became fully dependent on the magnitude of *competitive* prices in the market, and so as the oil rents (Bina 1985, 1992, 2006).

Finally, only after the establishment of the regime of "50–50 profit-sharing," which was directly dependent on posted prices (a mechanism of appropriation of the oil rents), did the oil-exporting rentier states become interested in the magnitude of the posted prices. The higher the posted price, the larger the amount of oil rent due to be appropriated by the oil-exporting nations, OPEC and non-OPEC alike. As we have already pointed out, the formation of OPEC prompted by a unilateral reduction of the posted price in the Persian Gulf oil region (see Tariki 1963, Alfonso 1966). This reduction led to a steep decline in the magnitude of oil royalty collected by the oil-exporting countries in the region. Similarly, there was a simultaneous reduction in the posted price of Venezuelan oil. And this was a trigger behind the joint protest by Venezuela and the other four Persian Gulf oil producers manifested in the first meeting of OPEC in the fateful September of 1960 (Rouhani 1971, Mikdashi 1966, Terzian 1985).

## Decartelization of Oil and Competition

Contrary to the US neo-cartelization era since 1911, the postcartelization era has been about the global commoditization of oil through competition. The price of oil, which once was the result of either the IPC's basing-point control abroad and the subject of the coordinated "market demand factor" strategy in the US domestic oil industry, now found a new footing in the budding spot markets around the world, as long-term colonial oil concessions were gradually yielding to the all-embracing and blind forces of the market. As the control of oil was about to be freed from the quivering clutches of the IPC, the dawn of a *new global oil order* was visible on the horizon; the reintegration of oil by means of global competition brought all oil-producing regions of the world under one pricing rule. Thus, over and above the general rate of profit, the more productive oil regions could appropriate differential profits commensurate with their differential oil rents. This has since been a universal rule for OPEC producers and non-OPEC producers alike. In this new milieu, at the various oil regions, the "posted prices" that were lingering from the IPC era were forced to be in sync with each other. More important, these "prices" also were compelled to mirror the actual market conditions heralded by the now spreading spot oil markets around the globe (see Minard 1980).

Finally, the commencement of the decartelization era coincided with the 1973–74 oil crisis. But as it turned out, far from being an exception, the periodic crisis is a regular feature of any globally integrated system of capitalist production and reproduction (Bina 1985). The oil crises in this new order are like a bellwether acting to announce the periodic restructuring of capital from time to time. They are the Guardian Angels of capital, so to speak. As was pointed out earlier, since the decartelization of oil, there has been a fundamental transformation in the nature of OPEC. OPEC has developed into the structure of global oil. The limits of OPEC oil rent are also set by this structure, far from the combined intentions of its individual members or the collective actions of producers in this body. And as has been implicitly or explicitly shown in this book, the size of oil rent is not determined by the action of OPEC but by the specific social and material forces that made oil a marvel of capitalism.

## Forms of Mineral Ownership

The entire history of OPEC bears testimony to the fact that there has always been a potential or actual conflict between the members of the organization and the controlling IPC since its inception in 1960 (Hirst 1966, Hartshorn 1967, Rouhani 1971, Alnasrawi 1985). To be precise, the formation of OPEC itself was the very consequence of such a conflict. The primary source of this antagonism, however, has been recognized in the contrasting relationships between the owners of the mineral deposits and those of the concessioners. For the exploration and production of oil may become intertwined with the imposing barrier of subsurface ownership. In the production of oil, the accumulation of capital may face a potential constraint from the ownership of oil reserves in the valorization process, depending upon the degree of capitalist development. The result is the necessity of formulating a conceptual framework that will be capable of capturing this additional characteristic that is closely related to the pricing of oil. The consequence is to develop a theory of economic rent consistent with the globalization of oil and, at the same time, suitable for both OPEC and non-OPEC producers (including the US producers) alike (see Bina 1985, Ch. 8).

Although it is possible for capitalist producers to have sole ownership of the mineral deposits themselves, the fact remains that the domain of oil production has been subject to oil contracts and leases between two socially and economically distinct classes, that is, capitalists and owners of oil deposits. In the former instance, the entire surplus profit will be appropriated by the capitalist producers (i.e., oil companies) themselves. In the latter case, given the interaction between the rentier and capitalists, surplus profits linked to differential productivity may turn into differential rents headed for the owners of mineral deposits. It is imperative, therefore, to recognize the dynamics of oil production in conjunction with the reciprocity of the structure of capital accumulation and the ownership of oil reserves. Moreover, in the oil industry, the barrier of reserves ownership is already *internal* to valorization and the accumulation process, both in the United States where the "rule of capture" is adhered to and in OPEC (and other non-OPEC oil regions) where the *state* is the sole and sovereign owner of the subsurface.

As has been argued in this book, in the United States where the rule of capture is observed, the fragmentation of oil leases has discouraged oil producers from engaging in additional oil exploration by way of extension or enhancing the capacity of oil reservoirs. Oil exploration may not seem commercially worthwhile, where it necessitates a large tract of land with a highly fragmented ownership of oil deposits and too many landowners with whom to negotiate. By the same token, the enhancement of oil production (secondary and tertiary recovery) may not be entirely possible, where the unification (or unitization) of the entire reservoir faces the obstacle of fragmented land and subsurface ownership. Thus, there are barriers to both capital deepening and capital widening in the US oil industry (Bina 1985, Ch. 7). These barriers together tend to influence the pace of capital accumulation and the amount of reserve addition in the United States (Miller 1973).

As for the state ownership of mineral resources, there are two interrelated aspects that must be recognized: the economic form of state mineral ownership, which is somewhat different from that of private ownership; and the political form of state ownership of mineral resources, which originates from the nature of the state and its institutional characteristics. The state ownership of minerals is different from its counterpart on the private side, in that it is not subject to the barrier of fragmentation of ownership. Instead, it contains the seeds of another potentially formidable barrier that stems from the state's role in the accumulation process, in addition to the collective representation of landed property. Hence, depending upon the nature of the oil-producing/rent-collecting states, global oil accumulation might be affected. While this may be a point of reference for the global production of oil in association with the rentier states, valorization of the landed property is a synthesis capable of overcoming any possible barriers. As Marx remarked so eloquently: "*The true* barrier to capitalist production is *capital itself*" (*Capital*, vol. 3, p. 358, emphasis in original).

This leads us to one more dimension of economic rent in the oil industry. The notion of rent is applicable to the private and state ownership of oil reserves in both OPEC and non-OPEC producers alike. In the meantime, the role of OPEC, as an organization of oil exporting countries, depends also on the appropriation of oil rent based upon its twin economic and political dimensions. OPEC is a rent-collecting agency whose effectiveness depends on the ability of its constituent members to capture the industry's surplus profits in the form of differential oil rents. Thus, the basis of OPEC's strength (or weakness) must ultimately reside in the global oil accumulation itself, which also serves as a powerful medium in the transformation of differential profits into differential oil rents.

## OPEC Profits and Their Limit

There are four distinct periods in the early history of OPEC: (1) the formative years, 1960–69; (2) the 1970–74 period of hasty transition; (3) the 1975–80 period of unsettled and slow adjustment to the global change; (4) and the post-1980 adjustment to the globalization of oil.

In the early formative years, OPEC was directly dominated by the IPC. In this period, even though global oil was moving toward the end of its transitional stage, the role of OPEC was passive and the magnitude of its oil rent was not even slightly reflective of the existing interregional differential productivities. OPEC members had no control over the production of oil or the conditions of contracts associated with the IPC (Cattan 1967b, Bina 1985, Ch. 3). In the transitional period of 1970–74, OPEC as a whole was a little schizophrenic, displaying divisions and detachment with respect to absorption and digestion of the enormous change that was speeding through like a runaway train. Aside from the inconsistent political posturing by the member countries, the rapid structural changes that had been long in motion, beneath and against the visible façade of US foreign policy—particularly manifested by the veneer of the Nixon Doctrine in the Persian Gulf—were too much for OPEC to bear in such a short time. Besides, not all the member countries were on a comparable level of economic development; neither were they on the same level of social and political development. Yet, against all unevenness and incongruity, the share of OPEC's oil rents had skyrocketed in this period, as the significant rise of "posted" prices between October and December 1973 would demonstrate (see Tanzer 1974, Stork 1975, Blair 1976, Bina 1985, and Alnasrawi 1985). The necessity of an active role on the part of OPEC, however, cannot simply be elucidated by such an active role itself, without entering the realm of tautology or circular reasoning. These activities are tantamount to both *necessity* and *sufficiency* of the objective forces that led to the unification and production of oil on a modern capitalistic basis. Thus, essentially, the quarrel of OPEC and the IPC ought to be understood in light of the internal contradictions of capitalism on a global scale, and not so much as the skin-deep "nationalistic" outlook of OPEC.

## Concluding Remarks

The development of OPEC has coincided with the period of transition of the oil industry from a cartelized global market under the control of the IPC, to the decartelization and competitive globalization since the mid-1970s. Here, basing-point pricing, long-term colonial concessions and gentlemen's agreements gave way to competitive spot-market pricing and the futures markets across the globalized oil industry. It may be a bit odd in practice in an ordinary business today to set different prices for the same commodity and to hope they be insulated for long in a competitive market. But it is more abnormal to set a lower price for the more productive oil and higher price for the less productive oil, and when it becomes apparent that isolating them from each other is not working, instead of cutting the price of the latter, one cuts the price of the former. This practice demonstrates that there was no market in an ordinary sense under the IPC; that is, a modern capitalist market, as a social arena that could validate the exchange of commodities free of the irrationality of the administered prices (i.e., cutting the wrong price) as the above example has shown. This is what the IPC did in 1959, which then led to the formation of OPEC in 1960. This transitional period characterized by the variation of oil rent grounded on the posted prices, a rudimentary form of rent that is compatible with the development of capitalist social relations in the Middle East, Latin America, and elsewhere outside the advanced capitalist countries. Such a development naturally brought with it the related contradictions and reciprocal relations that are particular to modern extractive industry. Thus, the major task of OPEC, which was set up primarily in response to the declining posted prices, must be seen as a consequence of the development of modern rent relations, and not the other way around. In other words, OPEC is not the cause but the symptom of the advancing modern oil rent relations across the member countries.

In a broader context, the story of OPEC reveals a profound theoretical impasse for neoclassical theory and mainstream economics as has been shown above. The literature on OPEC depicts a rather pessimistic picture, a kind that is steeped in a profound theoretical impasse, and normally reserved for a crisis of Kuhnian proportion. This impasse is primarily the corollary of a romanticized view of competition, which constitutes a point of theoretical departure in mainstream economics and, sadly, in some of its heterodox counterparts. There is a *difference*, in theorizing, between abstracting from complexities of reality and abstracting from reality itself. Orthodox economics, in general, and orthodoxy in OPEC literature, in particular, appear indifferent to this critical difference.

In sum, OPEC itself had gone through distinct developmental stages that were in conformity with the transition from the cartelization to decartelization of oil. The limit of OPEC oil rents is the magnitude of the interregional differential productivity. During the gluts in which the excess supply of oil from more productive regions causes the least productive oil regions to curb production, the magnitude of differential productivity (and differential profitability) is smaller than the period in which there are shortages. Thus, the magnitude of differential oil rents is lower in the period of persistent glut than in the period of persistent shortage. Accordingly, OPEC may seem "clumsy" during the glut and tactful during the shortage. OPEC is the collection of (rentier) oil-exporting countries not because of the alleged possession of "market power," but because of the substantial differential productivity of oil belonging to its membership, which, in global competition, provides substantial revenue in terms of differential oil rents. Therefore, the genie that Adelman (1980) was speaking of in reference to OPEC must be a fantasy—a genie of the dream. The real genie is the genie of globalization that has long been out of the proverbial bottle. It is undoubtedly obtuse to call OPEC a cartel—it is but obtuse.

# 4

# The Globalization of Oil

*Sweet are the uses of adversity,*
*which like the toad, ugly and venomous,*
*wears yet a precious jewel in his head...*

—William Shakespeare
*As You Like It* (Act 2, Scene 1)

## Introduction

In recent memory, it does not seem an overstatement to say that no commodity has been more frequently on the public's mind than oil. Yet oil appears to have remained enigmatic, if not ethereal, beyond the grasp of either the amateur or, for that matter, the self-styled expert. This mystical appeal could be attributable to the not-so-distant colonial history of petroleum that is still causing intrigue and captivating the public's imagination as the old collective memories are triggered by frequent, related events. But there might be another reason that is more intricate, more counterintuitive and consequently prone to misperception; and that is the complexities surrounding the interaction of capital and the landed property (in subsoil) in its truly modern and fascinating manifestation—the query of oil rent. Hence, the flimsy and fragmented view of oil, stripped not only of its complexity but of its authenticity and evolutionary history, also adds to all this mysticism. This lack of unprejudiced perspective is precipitated also by mainstream economic theory in its doctrinaire and irredeemable manifolds; in politics and social and ideological arenas; in public funding and public policy; and in the active lobbying and media coverage it receives. Thus public perception cannot be but held hostage. Sadly, the sweeps of captivity have not left too many stones unturned: left-leaning activists and intellectuals travel on the same route. Why should then oil be without mystery?

This chapter attempt to calibrate oil in its global setting toward demystification, and go beyond ghostlike political reminiscences that are still tiptoeing around the control of oil in today's postcolonial imperialism and posthegemonic America. We shall try to lay bare the specific grounds for oil economics and depict the evolution of oil stage by stage. In so doing, we also make an effort to establish a *synthetic* framework for the interaction of capital and landed property (in subsoil oil deposits) and to identify the dynamics of global oil rents. This approach has an urgent implication for two critical themes of our time: (1) the production of fossil fuels and the environment; and (2) the alleged connection between oil and war, and the real reason behind the recent US interventions. This theory concerns both premises. Yet, our focus in this book is to stress the second premise. As shall be demonstrated below, our theory and the nature of reality on the ground run counter to the tone of conventional *unwisdom* in many ways. For instance, the many liberal and radical leftists, who utterly submit to orthodoxy in economics, nevertheless show little enthusiasm for its policy prescription, for example, in the subtext of *drill, baby, drill* in Alaska's Arctic National Wildlife Refuge (ANWR); or in the case of US "dependency" on foreign oil and the subterfuge of "Project Independence"; or the views that had long been displayed in the media by right-wing economists (e.g., James Schlesinger, *Challenge Magazine*, 1991), and now dusted off and narrated verbatim by the liberals and leftists in the antiwar circles to fight against the alleged *war for oil* in the aftermath of the US invasion of Iraq (see Klare 2003, 2004; and Bina 2004a, 2004b for a critique).

The success or failure of any theoretical construct often relies on whether it is adequately grounded in a relevant, consistent, and transparent methodology. In this chapter, care will be taken to avoid the axiomatic, speculative, and mechanical approaches that exemplify the orthodox, and in many cases, heterodox economics of today. We attempt to steer clear of the ideal spectrum of the market-structure theory (and ideology) common to both orthodox and heterodox economics across nearly all related subfields, including energy economics and industrial organization. The investigation in this chapter initiates with the examination of pertinent historical records and then proceeds to understand the transformation of the oil sector across the world. And only then the task of conceptualization is carried out by identifying the milestones through which it has passed and the qualitative shifts in the political, institutional, and epochal context of these milestones brought to bear in the making of the modern oil industry we recognize today. We are interested in laying bare the reality lurking beneath the concrete, complex, and convoluted surface of this evolutionary phenomenon, namely, globalized and competitive oil, and going deeper than the skin-deep expressions and contingent actions of the agents involved. Therefore, the role of theory and thus abstraction—following the celebrated tradition in philosophy (and methodology), from Plato (circa 428–347 BCE) and Aristotle (circa 384–322 BCE) to Rumi (1207–73 CE) and Hegel (1770–1831 CE)—is to undress. This is called *real* abstraction. Thus a lucid and logical discourse on face value should go beyond the face but not beyond the value.

Such an *undressing* of the *concrete* also resonates with Karl Marx, as he cogently upheld that "all science would be superfluous if the form of appearance of things directly coincided with their essence" (1981: 956). Therefore, real abstraction is to peel off the overlaying concealments, layer by layer, and to strip the subject matter down to its bare bone—though not by crushing the bare bone to axiomatic dust generated by the artifice of imagination. Thus a *real* theory is but the outcome of a *real* abstraction. In other words, real *abstraction* has little to do with speculative axiom-mongering and more to do with stripping its (concrete) subject from the convoluted and contradictory contingencies to reveal the essence of its manifold dynamics. The prerequisite for all this is the specific knowledge of the subject in detail and in concrete historical transfiguration. This is how the subject of oil is treated in this book. The method of abstraction utilized in this book is not axiomatic (i.e., speculative); it does not intend to calibrate the authenticity of oil by inching toward an ideal theory by way of "successive approximation"; it does not embark on "logical positivism," "methodological individualism" or Cartesian typecasts; the theory presented here is precisely a result of *real* abstraction through mediation of the concrete (historical) facts on the transformation of oil—from a semicolonial/cartelized entity to the postcolonial (and competitive) industry of today.

The methodological issues then for us is to investigate the transition of oil from the previous era to the present postcolonial era (1) in the presence of enduring concentration and centralization of capital in global competition and (2) in the presence of landed property (in subsoil) across all oil regions. These two entwined issues bring us to gauging the evolution of oil within the identifiable stages toward the present, while also attending to the unruly question of rent and the interface of capital and landed property (of oil deposits) in the course of advancement toward globalization.

## THE STAGES OF THE OIL PRODUCTION

For our theoretical purpose, and from the perspective of the evolution of a modern industry, we divide the entire history of Middle Eastern oil into three stages of development: (1) the era of colonial oil concessions, 1901–50; (2) the era of transition and transformation, 1950–72; and (3) the era of postcartelization and globalization, since the mid-1970s. Given the early discovery of oil in the United States (1859), a slightly different, yet substantially overlapping, periodization may be applied to the US oil industry: (1) the era of classical cartelization and early oil trusts of 1870–1910; (2) the era of regulated neo-cartelization of 1911–72; and (3) the era of globalization, since 1974 (Bina 1985, Ch. 3). These historical stages are not arbitrary but, as a corollary, reveal the evolution of capitalist social relations in the world oil industry.

A close examination of the entire period from 1870 to 1970 reveals that predominantly administered pricing (i.e., *unmediated* accounting calculations) and cartelized practices were the rule. Such a framework, however, had begun to lose its effectiveness in the 1950s and 1960s, as proliferating market forces did overcome the Achnacarry networks of the International Petroleum Cartel (IPC) (Blair 1976: 80–90, Federal Trade Commission 1952).[1] The 1928 Achnacarry Agreement inaugurated a new era of cartelization since the US antitrust law of 1911, which had led to the breaking up of Rockefeller's Standard Oil Trust. This was in response to the worldwide irreconcilable price wars that were in full swing at the time when there was no adequately developed global oil (capitalist) structure that would objectively mediate and manage all this perpetual chaos into a forcible, regulating reconciliation. This time, the control of oil meant cartelization of oil under the tutelage of a leviathan, from ocean to ocean, across the entire geography of the world, minus the Soviet territory. Blair charmingly summarizes the seven sacred tenets of this infamous agreement as follows:

> Alarmed by the rapidity with which the price war has spread from India to America and then back to Europe, the heads of the three dominant international majors met at Achnacarry Castle in Scotland to prevent the recurrence of such disturbances. Walter C. Teague, then president of Exxon [Standard Oil of New Jersey], was quoted by a trade journal as saying, "Sir John Cadman, head of the Anglo-Persian Oil Co. [BP] and myself were guests of Sir Henri Deterding [head of the Royal Dutch-Shell] and Lady Deterding at Achnacarry for the

> grouse shooting, and while the game was a primary object of the visit, the problem of the world's petroleum industry naturally came in for a great deal of discussion." Referred to generally as the As Is Agreement of 1928 or the Achnacarry Agreement, the product of this discussion was a document, dated September 17, 1928, setting forth a set of seven principles and outlining in general terms the policies and procedures to be followed in applying them. The principles provided for: (1) accepting and maintaining as their share of markets the status quo of each member; (2) making existing facilities available to competitors on a favorable basis, but not at less than actual cost to the owner; (3) adding new facilities only as actually needed to supply increased requirements of consumers; (4) maintaining for each producing area the financial advantage of its geographical location; (5) drawing supplies from the nearest producing area; and (6) preventing any surplus production in a given geographical area from upsetting the price structure in any other area. The last point asserted that the observance of these principles would benefit not only the industry but consumers as well. (1976: 55)

Given the necessity of tight control and awesome task of administration, the cartel had to issue several supplementary memoranda subsequent to the original agreement. Blair writes:

> As the companies became increasingly familiar with the troublesome problems of trying to make a cartel operate successfully, the instructions had to cover a growing number of issues and at the same time become increasingly specific and precise. The principal topics with which they dealt were: (a) fixing quotas; (b) making adjustments for under- and overtrading; (c) fixing prices and other conditions of sale; and (d) *dealing with outsiders.* (1976: 57, emphasis added)

The first stage in the development of the Middle Eastern oil industry coincided with the rudimentary development of capitalism and absence of full-fledged modern landed property. The private ownership of land excluded the ownership of subsoil, including the ownership of minerals underneath. A typical oil concession included the surrender of the right to explore, develop, and produce oil, natural gas, and related substances to the concessionaire, an international oil company. And from both legal and theoretical standpoints this surrender of the right to explore, develop, and produce should not be confused with the surrender of ownership of the resource (i.e., oil deposits in place) to the contracting oil companies.[2] The term *concession,* rather than *lease,* refers to a contract between a private entity (i.e., a company) and a government (i.e., a would-be sovereign entity). The oil concessions during this first stage (1901–50) had more or less the following commonalities:

1. They nearly covered the entire subsurface of the land in a country or territory.
2. They had a long duration that normally extended beyond 50 or 60 years.
3. They were only a handful of cartelized concessionaires worldwide.
4. The terms of the concessions were uniform.
5. The principal financial obligation was the uniform payment of royalty.
6. The financial terms were extremely moderate.
7. There was little change in the terms and conditions of these concessions.

> The laws of the oil concessions [i.e., the colonial contracts] governing the *dominated* oil regions of the world, including the Middle East, are substantially different from the leasing contracts that prevail in the United States. It should be noted that the essential characteristic of the U.S. leasing practices stems from the structure of ownership of the subsoil, which is included as a part of the ownership of land. Due to the observance of the *rule of capture,* in the United States, the materials obtained from the subsoil belong to the owner of the land. (Bina 1985: 22)

Thus, from the beginning, capital investments in exploration, development, and production of oil had to come to terms with two separate systems of landed property in the subsurface across the globe. At the same time, from the standpoint of the stage of development, there emerged the tendency to a rudimentary valorization of landed property in these territories as opposed to a full-blown valorization in the United States (valorization of the landed property leads to the formation of rent, as a category, which in turn depends on the prior establishment of capitalism and capital as a social relation). That is why the industry as a whole—a disorderly conflation of different social relations in colonial and semicolonial settings—had to be managed by direct control and crude and unmediated cost and price calculations. Basing-point accounting, which is illustrated in a bit of detail in the chapter on OPEC, was essentially the main springboard of pricing in the period.

The second stage in the development of the Middle Eastern oil industry was the gradual objectification of market forces that eventually led to decartelization and abandonment of administered pricing of oil through the crisis of 1973–74. This stage saw the uneasy coexistence of the declining cartelized mechanisms and practices, and the rising proliferation of market forces that carried and conveyed the spread of competition against the prearranged production, captive oil concessions, "gentleman's agreements," and arbitrary accounting of oil royalties (and rents) according to fictitious "posted" pricing. Any transitional period, by necessity, tends to portray the amalgam of the vanishing past and the emerging future. The breakdown of the cartelization of oil was the consequence of certain evolutionary changes beyond the cartel's surrogate allocation and accounting system that had long been skillfully employed across the vast, untouched, and presumably passive geography of production. In one important sense, in contrast to its American counterpart, the history of the cartelization of international oil is a remarkable story of "primitive accumulation" (Thomson 1990). The cartelization of oil is indeed a prehistory of germinating capitalist social relations in these regions of the world—a prehistory of capital. Therefore, it would be a partial assessment if the focus of the analysis were to be merely on imperialism and outright plundering in this period. In addition to a more palpable issue of *nationalism*, the question of *class* that is often out of sight and lurking beneath all these occurrences must be taken to account in itself and as a prerequisite for organic unity of oil in globalization and thus the relevance of the *law of value* in the coming years. These two issues, although inseparable at the time, must be analytically dissected for the sake of the evolution of capitalism as an ultimate trump card and as a durable social relation, and the identity of imperialism as an unambiguous and identifiable period (i.e., a specific *epoch*) in the development of the former.

The transition in this second stage shows that the spread of capitalist social relations, via oil, was not only contradictory but also contagious. Historically, however, the triumph of cartelization sowed the seeds of its own destruction. Introduction of foreign capital in the exploration, development, and production of oil and the germinating capitalist social relations in many of these oil territories have eventually led to the valorization of landed property under capitalism. Therefore, this transitional stage is the beginning of the unraveling and dismantling of the ad hoc and fragmented accounting schemes that stitched the US oil basing-point system, at the Gulf of Mexico, to the newly devised (i.e., the cut-rate) posted prices at the Persian Gulf. This provided the companies with an opportunity to pocket not only the monopoly oil profits but also the lion's share of the oil royalties.

As we have expressed elsewhere in this book, some of the basic identifying features of this period are (1) the arbitrary division of oil profits and oil rents—starting with 50–50 profit sharing, (2) the elimination of "phantom freight" and the designation of a second basing point at the Persian Gulf,[3] (3) the nationalization (1951) and subsequent denationalization (1954) of oil in Iran, (4) the formation of the Organization of Petroleum Exporting Countries (OPEC),[4] and (5) the rise of independent oil companies and the demise of the Achnacarry (Alfonso 1966, Bina 1985: 21–35, Mikdashi 1972). During this period, given the desire for stabilizing the basing-point price of oil at the Gulf of Mexico, US domestic oil has also been controlled (Blair 1976: 121–203). This basing-point system, erected upon the wellhead price of US oil (at the Gulf of Mexico), was used as a universal (accounting) yardstick for the pricing of oil anywhere in the world (Federal Trade Commission 1952).

The so-called 50–50 profit-sharing was not really about sharing the oil profits equally between the oil exporting state and the IPC. This scheme was motivated by the dissatisfaction over the 1943 IPC-sanctioned oil laws that were ratified by the government of President Isaías Medina Angarita in Venezuela. These laws had given carte blache to a consortium of oil companies that either operated as the springboard of the IPC in Latin America or capitulated to the same cause. However, the Medina government imposed higher taxes on profits by foreign capital and manipulated the 1944 Venezuelan national budget in such a way as to show earnings equivalent to 50 percent of oil company profits. This raised eyebrows in the US State Department that was at the time behaving more zealous and acting "more catholic than Pope" in such matters. But the IPC was more relaxed and thought that as long as there was no violation against the 1943 laws, this measly (income) tax would be tolerable. The confidence of the IPC was not built on toleration alone, but on the solid-rock foundation of fabulous oil acquisitions and lucrative earnings, which were kept out of sight and guarded meticulously in an off-the-book fashion, and which would offset tens of thousands of times such nickel-and-dime propositions (see Painter 1986, Ch. 6: 128–34).

After the left-of-center government of Rómula Batancout took over and J. P. Pérez Alfonso was at the helm in all oil-related issues, eventually on November 12, 1948, and not without a fair amount of uphill battle and struggle, Venezuela was able to pass a new income tax law that forged "a 50 percent rate on any sum by which the company's net profits exceeded the government's share of the company's earnings" (Painter 1986: 133). A careful look at this "sharing" scheme, of course, would reveal that this is not a genuine 50–50 sharing of the profits, but a simple division of "earnings" that were formally kept on the books, "net" of tangible, intangible costs from exploration, drilling, pipelines to refinery runs, intracompany and intercompany transfers, discretionary costs attributed to the administration, and finally countless other categories of real or fabricated costs—all wonderfully hidden by design from the purview of oil exporting governments. Therefore, a more fitting description for this arrangement would a 50–50 window dressing. But there is no question that this was a better deal than the arrangement based on fixed royalty.

Venezuela's struggle under Alfonso toward "50–50 profit sharing," had also borne fruit in Saudi Arabia in 1950. Yet, in Iran, due to characteristic intransigence displayed by both the British government and the Anglo-Iranian Oil Company (AIOC), even subsequent to the nationalization of oil and the ensuing negotiations, there was neither enough wisdom on the part of the British nor adequate maturity on the part of the AIOC that the world they had been used to had already changed even within the IPC. In this period, the dogmatic attitude of the British government and archaic conduct of the AIOC as a government within the government in Iran is a shameful reminder that little had changed in Britain with regard to the postwar international polity. The following passage captures the disagreement between Britain and the United States on the question of 50–50 profit sharing for Iran:

> At a luncheon meeting [of early April 1951 in London] characterized by "tenseness" and "sparring comments," McGhee [U.S. Assistant Secretary of State] tried to persuade Fraser [Chairman of the Board of the AIOC] to consider current realities in Iran and be more forthcoming. Fraser was adamant, saying that McGhee's understanding of the situation was wrong and that there was no need to give Iran any concessions. "Fifty-fifty is a fine slogan, but it seems to be of dubious practicality," Fraser added. McGhee concluded that Fraser "had not yet learned." (citations from McGhee's memoirs in Elm 1992: 86)

As documented rather judiciously and with exceptional precision by Elm (1992), the rumored offer of 50–50 profit-sharing to Iran by the British government or by the AIOC was not an offer at all; Britain wanted nothing short of restoring the 1933 oil concession and reinstating the AIOC to its prenationalization status (Elwell-Sutton, 1955).[5] That is why a violent removal from office of an elected prime minster, who enjoyed unprecedented international support, in a sovereign nation was the only option. Mohammad Mossadegh

(1882–1967) was finally toppled in a violent CIA coup d'état in August 19, 1953, in Iran. A faint twinkle of colonial victory shone in Winston Churchill's cunning eyes. The tyranny of the containment of genuine nationalism (and messy democracy) though was the hang-up of his American counterparts (see Kinzer 2003). Hanging on by the thin thread of the Soviet threat (and catalyst of Cold War) was merely an afterthought of this badly choreographed tragedy. John Foster Dulles (secretary of the state) and his brother Allen Dulles (director of CIA) in the Eisenhower administration, who had a heavy hand in minute details, would chuckle at all this nonsense and at the naiveté and sophomoric reconstruction of the facts by historians (and by international relations specialists) who have long been perpetuating this myth. Those who have made a cottage industry of the Soviet threat and Mossadegh's removal may not have a clear idea about the art of theorization and distinction of *core* from *catalyst* in formulation of theory (see Gasiorowski 1987, Gasiorowski and Byrne 2004, Blake 2009). These scholars discounted the very fact that the 1951 nationalization of oil in Iran was a threat to the survival of the IPC and that the question at that *stage* was oil—and oil only. Our thesis acquires further solidity with the events, including US reaction, surrounding the formation of OPEC in 1960. It is instructive to know that the legacy of Mossadegh's nationalization did not expire with American (and British) coup d'état against his government. Mossadegh's gift of self-determination was received rather enthusiastically by the watchful eye of the world while he was in internment under the control and custody of the Shah's regime, on behest of the US government. In 1955, after a long debate, the Third Committee (known as the committee of 60) of the UN General Assembly adopted a resolution that supported the right of economic and political self-determination—including freedom of disposal of one's own national wealth and resources. The only three members opposed to this resolution were representatives from Britain, the United States, and the Netherlands. The 1955 text adopted by the UN General Assembly reads:

1. All peoples have the right of self-determination. By virtue of this right they freely determine their political status and freely pursue their economic, social and cultural development.
2. The peoples may, for their own ends, freely dispose of their natural wealth and resources without prejudice to any obligations arising out of international economic cooperation, based upon the principle of mutual benefit, and international law. In no case may a people be deprived of its own means of subsistence.

3. The states, parties to the covenant having responsibility for the administration of Non-Self-Governing and Trust Territories shall promote the realization of the right of self-determination in such Territories in conformity with the provisions of the United Nations Charter. (UN Doc. A/C.3/L.489 in Hyde 1956: 856)

Returning to the run-up to the formation of OPEC, as the new and bountiful discoveries of cheaper oil in the Persian Gulf region came to a fruition, the new oil had not only displaced the US markets to the west of Suez but continued also on the way to markets on the US eastern seaboard. Thus the regional oil markets adjacent to the Western Hemisphere were supplied with the oil from the Persian Gulf. This prompted the international oil cartel to cut the Persian Gulf posted prices to prevent the interregional flow of oil toward the US market, thus complying with the tenet of the 1928 "As Is Agreement" reached in the Achnacarry. Historically, the posted price at both Gulfs functioned as an allocating mechanism for transferring and disbursing crude within the worldwide networks of the cartel. Therefore, while cutting the Persian Gulf posted price reduced the flow of oil from this region, it also diminished the oil royalties for this region both in terms of the magnitude (per barrel) and the quantity of output.

As explained elsewhere in this volume, the founding of OPEC was a response to the continuous cuts in the posted prices by the IPC in the late 1950s. The posted price of oil was cut due to a combination of factors, such as the 1958 recession, expansion of Russian oil production, and the imposition of the 1959 oil import quota on the US domestic oil market, which was by far the largest in the world. The last factor, which was devised to discourage competition from the US independent producers, is indeed the tip of the iceberg of US government endorsement of As Is Agreement (the Achnacarry) at the expense of both the US domestic consumers and the royalty earners of the Persian Gulf oil region. This was, however, concealed by the US government under the convenient cloak of "national security." It is noteworthy to point out in passing that once the deception of national security—and the pretense of "strategic oil"— was concocted, the tensions between the Anti-Trust Division of the US Justice Department and the State Department over the violation of the Sherman Anti-Trust Act of 1890 and the pertinent antitrust law of 1911 subsided once and for all. This ingenious invention is only the tip of the blunder associated with the myopic, immature, and intransigent foreign policy of this period (see Blair 1976, Ch. 7). The mood against the founding of OPEC in the Western media can be captured in the following passage:

> After ten days of relative silence [...the Anglo-American] press began to attack OPEC openly. The *New York Times* mentioned the organization for the first time on 25 September 1960 and then only to call it an "international cartel". Expressing the feelings of the oil companies, the paper stated "[g]enerally, the oil companies are opposed to any such government cartel. They consider it impossible to establish a fair and workable program (of regulated production) and fear that the result, in the long term, would be withering away of market outlets."...The threat was clear. A few days later the paper returned to the subject, accusing OPEC of "an interference with the principle of free enterprise" and stating that "oil men here do not believe that the international oil organization formed in Baghdad provides the answer to the problem of stabilizing prices." (cited in Terzian 1985: 46)

Behind the scenes and in concert and choreographed with the media, the unofficial US foreign policy was the policy of status quo in line and indeed hand-in-glove with the basic tenets of the Achnacarry. This, for instance, can be seen from the US defensive attitude in failing to recognize OPEC for nearly a half a decade after its formation. The following passage from the 1964 US-UK Memorandum of Conversation, while shedding light on the role of the US State Department, also reveals the early idea of the countervailing "oil consumer grouping" against OPEC, long before the 1970s:

> We envisage, said Sir Geoffrey [Harrison, Britain's Deputy Foreign Secretary], that a confrontation on OPEC issues might take place in different ways. (1) We might find ourselves in a position...to support the companies. This would have many drawbacks, including the invoking of Arab nationalist sentiments [that] provide potential for Soviet meddling and create internal political difficulties in the countries concerned. Because of these fears, the Shah [of Iran] was prepared to get out in front in order avoiding [*sic*] enactment of sanctions at the [24 December 1963] Riyadh OPEC meeting. He, in fact, blocked sanctions against companies. (2) A confrontation might arise with the Western European consuming governments...if difficulties over OPEC should lead to an interruption in the supply...(3) A price rise could likewise provoke a Western European consumer combination to oppose OPEC. However, we incline to the belief that a rise in prices will come about in any event and the European governments will just have to learn to live with it...Mr. Kelly [U.S. Assistant Secretary of the Interior for Mineral Resources] expressed agreement in principle

> with everything Sir Geoffrey had said...We are also worried about a consumer/producer confrontation and there is a chance we might provoke this sooner than necessary...By focusing European attention now on Middle East oil problems we may stimulate European thinking on an oil consumer grouping to counter OPEC... *We wish to avoid a confrontation between OPEC and OECD* in 1964...Sir Geoffrey said he wished to reaffirm the joint position reached in the June [1963] talks on the desirability of maintaining a stance of neutrality and non-recognition of OPEC. (1964: 319–20, emphasis added)

Britain's inflated posturing and American naiveté toward OPEC turned out to be a flop. It took nearly six years for the US government to realize that it was virtually alone in nonrecognition of OPEC. Thus the belated US action by default:

> The U.S.-U.K. policy of neutrality and non-commitment towards OPEC detailed in CA-386 (paragraph 8) has not prevented the OPEC from obtaining recognition from international organizations, specifically the ECOSOC and UNCTAD, and Austria has granted diplomatic status to the organization and its personnel. *In light of these and other successes by the OPEC, the U.S. G[overnment] intends to review the present policy towards the OPEC* and consider if some other policy towards the organization might more usefully serve U.S. interests. (Ball 1965: 333, emphasis added)

Toward the end of the 1960s, there occurred, inter alia, three major developments that entirely undermined the cartelized character of the industry in favor of the rising objective market forces and spot oil prices globally. First, there appeared transformative macroeconomic changes in OPEC's relationship with the IPC; this was reflective of changes in the internal development and potential integration of the oil exporting countries into the world economy. Second, there emerged the proliferation of independent oil companies, which is a telling story about the internal turmoil and erosion of power in the cartelized system of Achnacarry (1928–72). This was a grand experiment on the so-called barriers to entry, and in retrospect it was settled unilaterally by the eventual collapse of the IPC in 1972. To identify some of these "independents," names such as Ashland Oil, Occidental Petroleum, Amerada Hess, Marathon Oil, Continental Oil, City Service, Sun Oil, Union Oil, Philips Petroleum, and Getty Oil come to mind (see Blair 1976, Bina 2012c). Finally, there was a considerable increase in the exploration and development costs of US domestic oil, the costliest in the world, in both per/barrel and absolute magnitude.

The latter, in turn, translated into a significant increase in the cost of US domestic oil production. At this time, a close inspection of the US oil fields revealed (1) considerable fragmentation of the new oil leases associated with the US domestic exploration activities, (2) sizable fragmentation of oil leases (i.e., the dispersion of royalty ownership) in the producing oil fields in need of unitization and application of advanced oil recovery, (3) the veritable decline of the US oil finding rate (oil reserves added per well), following the 1970 US production peak, and (4) significant increase in the cost of successive capital investments in the secondary and tertiary recoveries in the aged US oil fields (Bina 1985, 1988).

In the meantime, in the early 1970s, the Texas Railroad Commission abandoned the policy of market demand prorationing after nearly four decades since the discovery of bountiful East Texas field. As Blair (1976) articulates, the 1932 prorationing (or, as labeled rather artfully, "conservation") of Texas oil rights after the Achnacarry Agreement was a substitute for unitization of the fields (and the application of advanced recovery), which practically led to the destruction of billions of barrels of ultimate US oil recovery. On January 1, 1970, the US federal oil depletion allowance was reduced from 27.5 to 22.0 percent. On August 15, 1971, the Nixon administration instituted the first phase of price controls. On January 11, 1973, mandatory price control turned into voluntary control. On August 17, 1973, the Nixon administration imposed a two-tier price ceiling on domestic oil: old oil (produced at or below 1972 levels from existing wells) was to be sold at March 1973 prices plus 35 cents; new oil (produced above 1972 levels from existing wells and from new wells) was free of control. In 1972 the infamous 1959 oil import quota (a friendly gesture to the IPC in the name of "national security") was rescinded (Blair 1976: 152–86). This is the same program that triggered further cuts in the Persian Gulf posted prices and led to the formation of OPEC. Finally, there was the devaluation of the US dollar, first in December 1971 and subsequently in February 1973, respectively at 8.5 and 10 percent. All these developments transpired well before the October 16, 1973, and January 1, 1974, OPEC price hikes. On November 15, 1974, the International Energy Agency (IEA) was formed.[6]

Eventually, the grand cartelized network of Achnacarry was unraveled piece by piece during the transition period. The gentleman's agreement gave way to the tumultuous forces of the market. The lack of control over the increasing volume of oil outside of the cartel's network did the trick. The development of adequate capitalist structure in the oil exporting countries led to de facto valorization of landed property in oil. This in turn transformed the nature of OPEC, notwithstanding the Trojan horses of the golden years of Pax Americana within OPEC that desperately searched for a middle ground. The US domestic oil fields were rationalized; the global oil industry was reorganized and unified through the crisis; and the price of production of the US oil had become the regulating price of production for the entire industry worldwide. The world oil entered into the era of globalization with unified market prices, global differential oil rents, and plenty of volatility (Bina 1985, 1992, 1997; Bina and Vo 2007).

The 1973–74 crisis must be considered as the mirror of much larger manifold transformations, namely (1) the worldwide unification of the oil industry—from the lowest to the highest cost structure—under one pricing rule, (2) the de facto nationalization and concurrent transnationalization of oil against the IPC by the oil rentier states, (3) the decartelization of US oil and rationalization of the US oil industry, (4) the universal valorization of the landed property and competitive formation of global differential oil rents, (5) the transformation of OPEC from a rudimentary rent setter to a full-fledged rent collector, (6) the proliferation of global oil markets, abolition of posted prices, and formation of global oil spot (and futures) prices, and (7) the redundancy of the unmediated (physical) access, utopian self-sufficiency, and dependency on a particular oil region (Bina 1989b, 1990).

The era of cheap oil/expensive oil was over. The law of one price (underpinned by *regulating capital* in the US domestic oil) had become a universal rule for all oil across the board. Yet, in realpolitik, the deception of national security, via the allegation of dependency and demand for access, led to tough talks and threats against Pax Americana's favorite son, the Shah of Iran, by Henry Kissinger and to the panic plan of a Rapid Deployment Force by the Carter administration.[7] On the supposedly analytic front, the post-1970s geopolitics of oil had essentially centered on the traditionally fragmented quarrels over the de-Americanization[8] of oil and concern over US domestic oil production, consumption, and imports. And it took nearly another decade for the United States, OPEC, and the emerging world to realize that ultimately these epochal changes were irreversible. Decartelization of oil also cut the umbilical cord of the US foreign policy from oil. The IPC was in a variety of ways a beachhead with multiple economic and political outposts in the oil-producing countries. The companies within the IPC often operated as a government within a government in many of these countries. The Anglo-Iranian Oil Company in Iran was a notorious class by itself in this regard. This was probably as important, if not more, as the economic aspect of oil, particularly for the United States.

## THE THEORY OF THE OIL RENT REVISITED

The significance of oil rent and the necessity of its theorization are as old as the industry itself. Yet the direction of the Marginalist Revolution (the birth of neoclassical economics) has not been conducive to specific treatment of any rent, including oil rent. Moreover, in the span of several generations and the numerous contending exchanges, the neoclassical school eventually managed to do away with the specific treatment of rent. At the same time, despite the persistent pleadings by some rather prolific writers at the turn of the last century and beyond, the neoclassical paradigm and profession endorsed the tautology of "opportunity cost" and the simultaneity of general equilibrium for all types of production, including the ones that are valorized in conjunction with landed property.[9] Thus, rent was first generalized as the return on all "factors of production" before being euthanized and buried away from the myopic sight of the profession (Fine 1982b, Ch. 7; Hobson 1891). But the specter of rent has kept hovering over the spectrum of competition/monopoly, as a faint reminder. As is demonstrated in chapter 2 and again here in this section, rent is not merely a fleeting illustration but the pinnacle of the theory of value in its highest concretized form in Marx.

### Economic Rent: Valorization of Landed Property

The Achilles' heel of mainstream theory is nowhere more exposed than in the oil industry where oil rent is a crucial factor. There is no room for rent in the neoclassical framework, except in violation of idealized competition. There is also no specific rent in the general equilibrium framework where all the returns on the factors of production are rents. Yet, in partial equilibrium where the *specific* theory of rent is possible, neoclassical theory is only applicable to a static single-commodity world. That is why oil literature within neoclassical economics is profusely replete with the repetitive tautology of market power and monopoly.[10]

Being unable to utilize neoclassical theory to study the validity of oil rent, and reluctant to give up the reality (of oil rent) for the sake of neoclassical theory, we have had no choice except to return to the Ricardo-Marx literature on the political economy of rent as our prehistory (Marx 1968, Ricardo 1976). In the first chapter of Part 6 of vol. 3 of *Capital,* Marx lays out a framework for the meaning of rent in capitalist production. Marx identifies the pitfalls clearly:

There are three major errors that obscure the analysis of ground rent and are to be avoided in dealing with it[:]

1. [The confusion between the] various forms of rent that correspond to different levels of development of social production process... This *common character* of the different forms of rent... leads people to overlook the distinctions.
2. All ground rent is surplus value, the product of surplus labor... But the subjective and objective conditions of surplus labor and surplus value in general have nothing to do with the particular form, whether this is profit, or whether it is rent. They apply to surplus value as such, whatever particular form this may assume. They therefore do not explain ground rent.
3. A particular peculiarity that arises with the economic valorization of landed *property*, that is, the development of ground rent, is that its amount is in no way determined by the action of its recipient, but rather by a development of social labor that is independent of him and in which he plays no part. (1981: 772–75, emphasis in original)

It is important to realize that these conclusions are the result of Marx's complete theory of production, circulation, and distribution of value in capitalism. The first point is a caveat on the epochal identity of rent relative to the mode of production. The second point confirms that while rent is surplus value, the production of surplus value—as the effect of general conditions of capitalist production—has no automatic mechanism for identifying rent. Finally, the valorization of landed property (i.e., the materialization and magnitude of rent) is neither ad hoc nor undefined, notwithstanding the negotiation over the amount of rent between the land owner and capitalist investor. The magnitude of rent is determined by and consistent with the operation of the law of value. And, being the subject of valorization, the intervention of landed property is not an *antithesis* of capital but a part of the whole *synthesis* in valorization by the latter, which in turn is subject to sufficient development of the productive forces. This point is absolutely essential for correct interpretation and the precise meaning of absolute rent (AR) beyond arbitrary monopoly interpretations prevailing in orthodox as well as heterodox literature on rent.

Unlike Ricardo, Marx starts with the real experience whereby the least productive land will have to pay rent. He identifies this rent as AR. According to Marx, AR reveals the effect of monopoly of modern landed property on capital accumulation in agriculture; a monopoly that is fashioned in conjunction with the development of capitalist social relations. In this context, the monopoly of landed property amounts to a *synthetic* monopoly that can be prevailed over by the rapid pace of capital accumulation, measured straightforwardly by the organic composition of capital (OCC). According to Marx, OCC is an unambiguous measure of progress in agriculture (or in oil, in our particular instance) relative to all other sectors of the economy and, as such, depends very much on interindustry competition and the intersectoral mobility of capital. Thus, in the forceful process of *historical* transformation of values to the prices of production, some production prices remain above and some below the value contingent upon their corresponding deviation from the average OCC in the economy as a whole (Fine 1986, Shaikh 1977, 1990, Saad-Filho 1993). This implies that AR is necessarily subject to interindustry competition; and, consequently, its relevance as a category depends on the relative pace of capital accumulation within the sector in question. Thus, it is incorrect to depict AR arbitrarily as monopoly rent (Fine 1979). Parenthetically, this also alerts us to the inconsequence of the neo-Ricardian/Sraffian literature on the transformation from value to prices of production in Marx. Why the lack of consequence? Because this transformation is about a process that is *historical*—so as the *theory* that seeks to expound it. Additionally, the traces of this historical process can be substantiated from the prehistory of social relation of capital in major capitalist countries, where the intersectoral competition had not been developed beyond a rudimentary and undeveloped form. This is known in traditional jargon as economic development.

## Land as Capital versus Land as Landed Property

There is another tricky issue that is not new but often overlooked in the contemporary literature on rent theory; this is the necessity of distinction between land as *capital* and land as *landed property*. This question is an offshoot of Ricardo's principal distinction between profit and rent—an issue that compelled Marx to praise the classical school for its scientific part. And it is precisely the same question that early on had a critical role in the development of Marx's theory of value in the presence of landed property. The distinction between land as capital and land as landed property is critical for the identification of rent as a sui generis category; and Marx had been keen on it as early as 1847, when he composed *The Poverty of Philosophy* to respond to *The Philosophy of Poverty* by Pierre-Joseph Proudhon—considered by certain circles (posthumously) as the father of Paris Commune. This distinction, which separates the domain of interest-bearing capital from that of rent, is *categorical* in Marx. We read:

> The representative of land as capital is not the landlord, but the farmer. The proceeds yielded by land as capital are interest and industrial profit, not rent. There are lands which yield such interest and profit but still yield no rent. Briefly, land in so far as it yields interest, is land capital, and as land capital it yields no rent, it is not landed property. Rent results from the social relations in which the exploitation of the land takes place. (Marx 1973b: 143–44)

This passage provides consistency and theoretical punch for a coherent rent theory in terms of (1) the mutuality of modern landed property and its creator—capital, and (2) the methodological distinction between the two. Accordingly, blanket statements that often blur the difference between fictitious and interest-bearing capital, on the one hand, and rent, on the other hand, are out of step with Marx's methodology and out of sync with a coherent theory of rent in the present stage of capitalism.

Harvey (1999 [1982]), however, does not seem to appreciate this simple but subtle point. Here is what he had recorded in a more systematic volume of his work:

> When trade in land is reduced to a special branch of the circulation of interest-bearing capital, then, I shall argue, landownership has achieved its true capitalistic form. *Marx does not reach this conclusion directly*, although there are various hints scattered in the text to suggest that land-trading could indeed be treated as a form of fictitious capital (Capital, vol. 3, pp. 805–13)...But we ought to also the historical process whereby landed property is *reduced* to such a condition [i.e., as fictitious capital or "pure financial asset"]... *Monopoly power* over the use of land—implied by the very condition of ownership—*can never be entirely stripped of its monopolistic aspects, because land is variegated in terms of its qualities of fertility, location, etc.* (1999: 347–49, emphasis added)

The above paragraph tells volumes about Harvey's mangled and messy methodology on a myriad of subjects, from the assessment of contemporary capitalism to the judgments passed on the theoretical contributions both by classical political economists and their grand and all-encompassing critique in Marx's work. This is a glimpse of the scholarship, an enigma, by someone who likes to call himself a Marxist. To be sure, the alleged scattered hints in *Capital* vol. 3, Ch. 39: 805–13, that Harvey is alluding to are but the *reduction* of rent to fictitious or speculating capital. On the contrary, in the subsection, pp. 806–811, in reference to George Opdyke's *A Treatise on Political Economy* (1851), Marx speaks of postcolonial United States and points out that "the colonial states found on the basis of the modern world market" must be distinguished from "those of earlier times, and particularly those of antiquity" (p. 809). The context here is the *regulation of the price of untilled land by the price of the tilled land*, which in turn is determined by the capital investment in the tilled land. The latter part of this paragraph covers a revealing conclusion that has failed to catch Harvey's attention. The passage that hopefully may clear the deck on Harvey's hasty reading of Marx is as follows:

> Since, with the exception of the worst land, all types of soil bear rent (and this rent...rises with the amount of capital and the corresponding intensity of cultivation), a nominal price is thereby formed for the untilled portions of land, so that these too become a commodity, a source of wealth for their owner. This explains at the same time why the price of land rises for the entire area, even for untilled land. *Land speculation*, e.g., in the United States, *depends on this reflection* which capital and labor cast on untilled land. (*Capital* vol. 3, Ch. 39: 807, emphasis is added)

It is worthwhile to shift gears and in passing present a glimpse of Marx's theory of competition. Marx depicts competition as the antithesis of feudal monopoly and capitalist monopoly as "the negation of feudal monopoly, in so far as it implies the system of competition...Thus [he argues] modern monopoly, bourgeois monopoly, is a synthetic monopoly, the negation of negation, the unity of opposites" (1973b: 131). For Marx, and for Schumpeter, concentration and centralization of capital are the necessary ingredients of capital accumulation, and this constitutes ammunition in the competitive war of capital upon capital (Schumpeter 1942, Ch. 7; also Shaikh 1980). Likewise, integration in Marx (and Schumpeter) is not the antithesis of competition but its synthesis. And what is called "barrier to entry" is the very reflection of the continuous increase in the size of *regulating capital* in the battle of competition. Here, neither the fiction of pure competition nor the tautological construct of atomistic markets has any relevance to *real* competition in capitalism. If the capitalist concept of monopoly is synthetic, then it must not be mistaken for the orthodox notion of monopoly (see also Weeks 1981a, Ch. 6, 2011). Similarly, the monopoly of landed property must be treated in the same manner as *synthetic*, that is, by way of negation of negation.[11] Differential rent (DR), on the other hand, captures the effects of the variation in the quality of land together with the variation of capital investment in agriculture. Thus Marx elucidates:

> The level of rent, reckoned per acre, thus grows... as a result of increase in the capital invested on the land. And this takes place moreover with production prices remaining the same, and irrespective of whether the productivity of the extra capital remains the same, decreases or increases. The latter factors modify the degree to which the level of rent per acre grows, but not the fact that it does grow. This is a phenomenon that is peculiar to differential rent II and distinguishes it from differential rent I... The more the capitalist mode of production develops, however, the more the concentration of capital on the same areas increases, so the rent per acre rises... This difference in the levels of rent could thus be explained neither in terms of a difference in the natural fertility of the land types nor in the amount of labor applied, but exclusively in terms of the different kind of capital investments. (1981: 830–31)

Marx certainly takes his lead from Ricardo. Yet, his concept of rent sharply departs from Ricardo's in two important respects: (1) that the assumption of "no landed property" in Ricardo's theory is faulty and (2) that Ricardo's rule concerning the order of cultivation from higher to lower quality land has no support in reality. Marx's classification of DR into the two types of DR I and DR II corresponds to the application of an equal quantity of capital to equal-size lands of different quality and the application of different quantity of capital to a given quality land under cultivation respectively. The combined effects of these differential rents (the former arising from natural fertility and the latter from successive application of capital), however, do not lead to a linear separation (Fine 1979; see also Fine 1990, 1994). This point is crucial for Marx's theory of value for two reasons: (1) unlike Ricardo's, Marx's rent theory does not arise axiomatically from any generalizable, natural condition but from the historically specific valorization of landed property—hence no general theory of rent, and (2) the concurrence of normal size capital and least productive land affects the regulating price of production in agriculture.

The second point is critically important for our own specific theory of oil rent in which the presently least productive oil deposits should not be necessarily deemed as the least productive in their original natural state; once considerably productive, these deposits have been turned into their present state by successive investments of capital. Finally, AR is not a stand-alone concept, separate from DR II. Since DR II sets the limits of AR in the course of the landed property's valorization in the presence of interindustry competition of capital—manifested by organic composition of capital—this shows that Marx's theory of value (and prices of production) unites the process of production, exchange, and distribution, before ascending to its pinnacle of *theoretical* concretization via the theorization of rent.

## Valorization of the Oil Deposits

At the outset, we need to identify the system of landownership, hence the ownership of oil deposits, in the oil industry before attempting to address the question of worldwide valorization of landed property in oil. As pointed out earlier, there exist two separate systems of ownership rights in the oil industry: (1) the US rule of capture inclusive of the private ownership of subsoil and (2) the public ownership of the subsoil in the rest of the oil-producing regions. This, of course, presents us with two different forms of the appropriation of nature prior to valorization as landed property in production. As has already been argued, any investigation into the question of rent must be specific, given the specificity of the landed property involved: hence the specificity of the production of oil. Therefore, Murray's (1977, 1978) proposition that agriculture must set the alternative use of land for oil rent is a fallacious claim for two reasons: (1) the contention of alternative land use is acting as a short step away from the pitfall of tautological reasoning and thus preparing the ground for acceptance of the feeblest feature of mainstream theory, "opportunity cost" and (2) the analysis of oil rent has neither historical specificity nor the slightest bearing on the landed property in oil (see Murray 1977, 1978, and Fine 1983 for a critical alternative). But the most damaging aspect of Murray's (1978) contribution to rent theory is the invention of "founder's rent" and misinterpretation of Marx's AR. To be grounded in solid foundation, AR must be dynamically connected to the transforming forces of production in the valorizing synthesis of capital and landed property—a synthesis that is suffused in competition. Murray's "founder's rent" is a mental construct, a mode of appearance, which has no wherewithal for such a herculean undertaking. Yet this egregious error is praised by Ernest Mandel in his Introduction to Marx's *Capital* vol. 3 as follows:

> In the second part of his remarkable study "Value and Rent" (*Capital and Class* Nos. 3 and 4), Robin Murray makes the point (pp. 13 ff.) that settlers overseas could generally expect a "founder's rent" similar to Hidferding's founder's rent of large oligopolistic enterprises. I think he is right, at least with regard to overseas countries with above-average fertile land compared to West Europe. But he gives excessive weight to such "rent" in explaining international migration, capitalist expansionism and the origins of imperialism. (Mandel 1981, Introduction to Capital vol 3: 65, ff. #95)

The lack of appreciation for Marx's AR also appears in Mummer (2002), where the author makes up the term "customary ground rent." Aside from phraseology and context, Mummer's belated attempt also does not fare any better than Murray's (1977, 1978) original vintage. Both versions are impish disguises of monopoly rent.

There is yet another attempt at chasing away the ghost of Marx's AR and bringing back the good old orthodox notion of monopoly. Harvey (2010) offers a new version of monopoly rent as the "reserve price." Harvey (2010: 81) declares: "because, frankly, I do not think it [i.e., Marx's AR] works." To grasp the essence of Harvey's "reserve price" the following passage is instructive:

> The very existence of this reserve price testifies to the monopoly rent that attaches to all form of property rights claims under the institutional arrangements that characterize capitalism. Any holder of a property right can withhold access to that property, and refuse to release it until a reserve price is reached . . . In the case of oil wells, however, we are here dealing with a non-renewable resource, the reserve price on which is given by conditions of relative scarcity. Differential rents on oil wells . . . here shades into monopoly rents, as has so obviously been the case with OPEC's control over the release of oil into the world market at a rate which maintains or stabilizes prices at a given level. (Harvey 2010: 81–83)

The above passage speaks for itself, in light of the extensive analysis given in chapter 2 and elsewhere in this book; we have no need to reinvent the wheel again and again. The reader may decide for oneself with respect to Harvey's peculiar reading of orthodox economic jargons (notice the "reservation price") and grafting them onto his self-styled Marxian approach. In his new writings, Harvey also casually trespasses on subjects that he appears to have little knowledge—OPEC and oil are his recent forte (see Harvey 2003, Chs. 1–2, Harvey 2010: 77–84). Harvey's scrambled method in connection with so many pertinent theoretical, historical, empirical, and cross-disciplinary issues of our time is both perplexing and discouraging at the same time, particularly for someone like the present author who admires Harvey's courage for standing up on issues that matter, both politically and morally, and have long been among his sympathetic colleagues around the world.

The crucial question is whether the landed property in oil—that is, this specific form of landed property—receives AR and, if not, why not. The answer depends on the pace and development of capital accumulation in the oil sector as a whole, which in turn relate to the mobility of capital between the oil sector and the rest of the economy, signifying the extent of interindustry competition, measured by organic composition of capital in the oil industry. In other words, although AR is not a monopoly rent in an ad hoc neoclassical economic sense of the term, it nevertheless may tend to impede the inflow of capital from other sectors, thus obstructing the interindustry competition of capitals and exhibiting a below-average organic composition of capital. That is why it would be essential to make a distinction between absolute and differential rents and thus be able to demonstrate that the decartelization of oil (since the 1970s) corresponds with the development of differential oil rents through worldwide competition. This distinction alone is a critical step to understanding the complexity of the contemporary oil sector in terms of its competitive unification and globalization.

The fact that the US oil region has been heavily explored and intensively drilled is an indication that the valorization of US landed property (in subsoil deposits) under the rule of capture has been achieved with high organic composition of capital. This also goes for the lesser explored and more productive oil regions under the rule of public property. Two conclusions are in order here: (1) AR is not a monopoly rent, and (2) there is no AR in the oil industry. However, in comparative static terms, one might argue that the least productive lands (or subsoil oil deposits) will not be leased unless they receive rent.[12] But the least productive US lands (US domestic oil fields) are not necessarily the ones that are presently leased for exploration; indeed the least productive oil fields are the ones that are already heavily producing. These oil fields that were once deemed productive have now been turned into their present least productive classification by the successive application of capital. These are the kind of oil fields in which price of production meets the requisite of combination of the least fertile deposits and normal capital, thus forming the regulating price of production for the entire industry. Therefore, the rent of the newly leased lands (oil fields), in the above example, may not be an absolute oil rent but a differential oil rent.

Given the range of oil regions in the world, from the most to the least bountiful, the central issue is the formation of differential oil rents across the globe, which is subject to intraindustry competition. This presupposes the globalization of the industry, formation of differential rate of profit, and a single worldwide market value for oil. Similarly, we need to distinguish two separate forms of differential rents in the production of oil: (1) differential oil rent type I and (2) differential oil rent type II.

Because of the fact that the separate effects of DR I and DR II cannot be known in advance (i.e., the impossibility of separating the effects of worst land and normal capital in advance), an a priori application of rent theory cannot produce a meaningful outcome for our purpose (Bina 1992). Therefore, speaking of DR, by identifying its two forms (DR I and DR II) in abstract, neither provides any determinate solution to the question of oil rents nor allows for specific conditions that are pertinent to the dynamics of capital accumulation in the oil industry. Indeed, among the reasons for the rejection of a (catchall) general theory of rent in Marx's work is the necessity for distinctiveness and singularity of each type of landed property and the specificity of its valorization via-à-vis capital. That is why it was necessary to engage in a systematic study of the oil industry by taking up a posteriori the theorization of structure, institutions, and evolution of oil, in terms of decartelization, proliferation of the spot and futures markets, and the formation of differential oil rents across the globe in competition (Bina 1985, 1989b).

In US domestic oil, given the rule of capture, the fragmentation of oil leases, particularly when the size of the reservoir is huge, has long been troublesome for satisfactory unitization of the oil fields for secondary and tertiary oil recovery. These reservoirs, while they had been fully productive (i.e., nonmarginal) in their original natural state, having been subject to several rounds of successive investment, have declined appreciably in the process of advanced recovery. Hence, the successive application of capital, particularly to these larger reservoirs, has led to decline in natural conditions of the oil fields and, consequently, to their productivity decline. As Bina (1985) has shown, throughout the 1960s, the US oil capital expenditures (per barrel) in exploration, development, and production had risen remarkably. These are the old oil fields within the lower 48 states in the United States, which have been heavily under production for a long time. It is the individual production price (cost-price plus average profit) of these oil fields that is the largest in magnitude and thus set the regulating price of production for the US domestic oil as a whole. In turn, it is the production price of US oil that regulates the production price of oil in the entire world, which in turn operates as a gravitational center for the market price of oil anywhere in the world. These US domestic oil fields were also the epicenter of the crisis, which prompted the restructuring of the US oil industry along with other oil regions of the world in the 1970s, and which resulted in the worldwide formation of differential oil rents (Bina 1989b).

## Nationalization, Globalization, Volatility, and Ghosts of Yesteryears

One of the significant corollaries of the 1973–74 oil crisis was the en masse but gradual nationalization of oil that either de facto or by intent had taken place across the entire geography of oil under the IPC's concession system. These oil nationalizations were carried out by both the states that had an unfavorable view of US foreign policy in the region and those that thrived on such policies. The latter waited by delaying and opting for "equity participation," and even then it did not dawn on them that these de facto nationalizations are indeed an inbuilt mechanism and a gateway to the globalization of oil. The process started piecemeal in Iraq (1972–75), Libya (1971–74), Kuwait (1976), Saudi Arabia (1976), Venezuela (1976), and Iran (1979)—the last was de facto upon the fall of the Shah. The concession system was finally buried along with the IPC by the mid-1970s. The succession and frequency of these nationalizations should not be lost on an observant reader who grasps the very fact that globalization is not incompatible with nationalization of oil (deposits), as the two are embodied within the synthesis of capital and the subsoil property in competition. On the recent (re)nationalization of oil in Argentina see Bina (2012b).

The era of post-1970s globalization of oil was beset with recurrence of volatility and, by implication, widespread uncertainty across the world. Oil is a commodity whose point of origin is insignificant once it arrives in the interconnected global pool. Therefore, notwithstanding the differential regional costs of production, the market price of oil is universally the same. The short-run market price of oil is determined by the spot and futures oil markets. This means that, from the standpoint of globalization, there is no distinction between "cheap oil" and "expensive oil." The spot markets reflect the daily delivery of oil on a competitive basis. Regardless of the individual conditions at certain localities, spot markets transmit signals as market-clearing prices that set the short-run price of oil anywhere in the world. The futures price, on the other hand, refers to competitive delivery of oil, sometime in the near future, hence the likelihood of speculative bubbles in these markets. These prices are also a reference point in the long-term contracts.

Spot markets in NYMEX (New York) and IPE/ICE in London are the ones that for all intents and purposes set the daily price of oil globally. The benchmark for NYMEX is the West Texas Intermediate crude, while IPE/ICE trades in Brent Crude—the crude from the North Sea. In turn, the OPEC oil basket itself (a composite of spot oil prices of member countries) takes its cue from Brent (and, by implication, from the West Texas Intermediate crude), and thus consistently varies according to fluctuations in these competitive global oil markets. This concrete reality provides us with three interrelated theoretical points in the economics of oil: (1) that the short-term global price of oil does not necessarily depend upon the concrete, market-clearing (physical) equalization of oil demand and supply at each and every single location in the world; (2) that the converging pattern of long-run (random) market fluctuations is neither independent nor a cause of the long-run "production price"; and (3) that such a pattern ordinarily reflects short-run fluctuations (of demand and supply) about the gravitational center of the long-run price. This center of gravity is none other than the "production price" of the costliest oil deposits in the world. The last point elucidates that the production price precedes the formation of market prices for all oil globally.

The futures market (i.e., NYMEX, New York) is a hedge market that normally operates alongside the spot market. However, given its purpose, this type of market is not without speculation. Hence, the question is how much activity in this market is aimed at effective hedging and how much geared toward speculation. This is how the issue of "selling oil that you don't have" had transpired in the summer of 2008, hinting at speculation on Wall Street by putting down, say, 6 percent of the barrel, and turning around to sell 100 percent of the same barrel even when it is not yet owned. In this particular instance, the culprit is the lack of adequate regulation, combined with the drumbeat of war by the Bush-Cheney administration against Iran in the very same period.

On another pertinent issue, the lack of rapid response to the increasingly excess demand by the price relates to two conundrums: (1) the requirement of long lead time for building new capacity in the presence of market volatility and uncertain future prices and (2) the dilemma of switching off from shut-in capacity and back, without sustaining a considerable economic cost due to the loss of technical efficiency and possible damage to the reservoir. This situation is worse in the case of excess supply. The levels of shut-in capacity have already been set normally in advance in the majority of oil fields, including those that regulate the global price of production. These regulating oil fields are particularly hard-pressed in the case of declining prices. The plugging of oil wells is one (costly) option. However, once they are plugged the oil usually will be lost forever. To avoid damaging the reservoir, another option is to keep operating the oil fields and hope for better market prices tomorrow.[13] That is why the regulating price of production does not decline immediately unless there is a prolonged excess supply, in which case the levels of risks and losses are too great for these producers to continue.[14]

This situation may indeed trigger an oil crisis, leading to a worldwide restructuring of capital, a new regulating price of production, and the corresponding market prices within the global oil industry. Thus, the characteristic of such crises must be explained from within, that is, from the standpoint of the oil industry's internal dynamics, not according to the circumstances arising from the external contingencies. And oftentimes relying on *power* as both the premise and the end result (such as market power or political power) further mystifies the subject that has already attained the highest degree of complexity in the contemporary economy, polity, and international relations.

In the remainder of this section we need to clear up two additional, yet interrelated, issues: (1) the alleged cartelization of oil even after the oil crisis of 1973–74 and (2) the credibility of conspiracy theories. On both of these points that potentially feed on each other and that perhaps may have possible implications for the questions, such as US alleged hegemony or US intervention in Iraq, it is worth quoting Fine and Harris at length, via two separate paragraphs. The authors simultaneously illuminate and obscure the very essence of the 1973–74 oil crisis as follows:

> *If we now put aside the oil crisis of the early 1970s* and examine its *results,* we can see how the oil industry discovered a solution to the erosion of the world cartel and the pressures on domestic U.S. production. The large increases in the price of oil have sustained the profitability

> of producers in the U.S.A. and have guaranteed sufficient revenue in the world production *to bind the majors and non-majors together in a cartel that now includes both.* The result of this has been to create enormous surpluses on the production of oil from those reserves, nearly all, that are less costly to exploit than those in the U.S.A. What…OPEC nations and other countries have been able to do is to appropriate some of those surpluses. *That they can do so is a result and not the cause of the oil price increase.* (1985: 86–87, emphasis added)
>
> To some extent, *this might read like a conspiracy theory of the oil price increases* in which the latter was a solution to the problem of the industry. *Certainly, such a possibility should not be discounted* and such theories abound in discussion of the oil crisis. Some argue that the crisis was a U.S. device to improve its competitive position relative to its industrial rivals by forcing a high price of oil upon them, others that it was a device to improve the U.S. balance of payments position through the recycling of petro dollars. These may or may not have been the effects or the intentions of the actions of the various agencies involved, but the solution to the industry's problems came about through a definite process that can be recognized. (1985: 87, emphasis added)

Fine and Harris correctly point out that the rise of OPEC surpluses is the effect of larger surpluses that have emerged in the global industry. Yet, they incorrectly treat the emerging competition in the post-1973 decartelization of oil as a newly formed cartel that binds "the majors and non-majors together," despite the very fact that the era of price fixing, deliberate division of the international markets, and unmediated control of production is over. By focusing on the consequence of the crisis on US oil production, Fine and Harris point out that *"this might read like a conspiracy theory* of the oil price increases in which the latter was a solution to the problem of the industry." The phrase "might read" here, however, has a methodological connotation for the phenomenon of conspiracy that precisely points to the mode of appearance of the (raw) concrete and thus cries for *real* abstraction and dialectical (informed) appropriation. Yet, conspiracy hypotheses often act as the premise and as the end result at the same time and thus are not capable of overcoming their own tautological status.

It is therefore instructive to ask why the "majors" should knowingly conspire against their own interest, particularly when it comes to the surrender of their control over the bulk of world oil reserves and, not to mention, their incomparable position in the worldwide pricing of oil. And, more importantly, why would the majors (or nonmajors) be interested in making the domestic US oil production free of "pressures," when indeed such a relief will have to be eroded soon through the global restructuring of the entire industry in which a significantly higher magnitude of the differential oil rents will become the norm also for the so-called new US oil? In other words, why in the world would the old cartel (the majors) wish to swap their old exclusive position with a "new cartel," shared with nonmajors? Is it not the forces of real capitalist competition, which have been gathering strength all the while through the transitional period of 1950–72, that ultimately led to the grand implosion of 1973? Does the phrase "the cartel that now includes both" not further fuel the spread of confusion in the minds of some scholars, who tend to blend the notion of monopoly with hegemony, and encourage them to falsely rely on the "character of monopoly in the industry and an interpretation of the role of the dominant state in the oil sector, the United States?" (Bromley 1991: 58) And would this very observation not invoke the ghosts of old conspiracies once again?

Finally, on their face value, conspiracies are not suitable candidates for objective empirical verification reflective of the mediating institutions. The illusion of conspiracy, like the reflection of mirage, depends upon the real (mediating) material basis that is beyond the realm of conspiracy itself—shown in Bina (1985, Ch. 2). However, in the case of the International Petroleum Cartel under the Achnacarry (1928–72), the perceived coincidences of conspiracy, while necessary, obtains sufficiency by the very existence of the unmediated cartel itself. Hence Achnacarry, due to its mission and administrative nature, was a giant conspiracy onto itself. In other words, cartels and conspiracies are a complimentary aspect of a phenomenon that drives its livelihood from the lack of mediation and mediating institutions. Therefore, mislabeling and mischaracterization of the post-1973 decartelization (and globalization) of oil should alert us to double trouble and indeed a double misunderstanding. And, it is within this context that with all due respect we are asking: what is the assertion "*such a possibility should not be discounted*" supposed to mean?[15]

## CONCLUDING REMARKS

Capitalist social relations act much like volcanoes or hurricanes. Once they have gathered strength and become a formidable force, they devise their own patterns and externalize their own mechanisms of enforcement. The post-1974 globalization of oil is not an exception to this rule. The crises that followed in the late 1970s, mid-1980s, and beyond owe their *epochal* identity to the oil crisis of 1973–74, in which the valorization of landed property in oil obtained a global dimension. Consequently, the past four decades of oil production must be carefully distinguished from the prior 11 decades, before the oil crisis of 1973–74. This distinction must not be made on the basis of quantity alone but quality and the incidence of epochal change that led to decartelization and globalization of oil, and the subsequent fall of the Pax Americana. The implosion of the IPC and decartelization of oil had a profound effect on the foreign policy of the United States. It cut the umbilical cord of the policy that had long been enmeshed with cartelized petroleum as not only an economic godsend but, more importantly, grounds for overwhelming political control. And the collapse of the cartel was a double body blow to the Pax Americana itself.

As has been revealed, globalization of oil is manifested in the worldwide unity and contradiction of all oil regions in global competition, as the oil spot prices are its momentary reflection. Globalization of oil reveals worldwide valorization of the oil deposits manifested in the formation of differential oil rents across the oil regions of the world. Thus, the capital investment on the least productive oil region tends to regulate the *production price* of oil for all regions across the globe. The latter acts as a gravitational center (i.e., a price in the long run) around which short-run market prices would oscillate according to the laws of demand and supply.

The production of oil from the least productive oil region is entitled to a competitive profit. This reflects a normal rate of return on capital investments that, notwithstanding the risk and uncertainty, move rather competitively in and out of the industry on a regular basis. By comparison, oil production from more productive oil regions is entitled to a differential oil rent, in addition to normal profit. The long-run price of oil is set by the US oil production price (the least productive oil region), which in turn represents the gravitational center of short-run fluctuations of oil prices worldwide.

This has a profound implication for the question of environment, an issue that does not square with the alleged US "energy independence" and the desire for "self-sufficiency" in a global economy that has long been interdependent. But in a world in which producing and pricing of oil would no longer yield to the national boundaries, the question of energy independence is not unlike the idea of the tooth fairy in children's imagination. The problem becomes further complicated when the desire for energy self-sufficiency is measured with the reality of capitalism and capitalist markets in which nearly everything is antithetical to self-sufficiency. But strangely enough, the proponents of "self-sufficiency" have no concern about other commodities but oil. This is precisely how the ideology of "drill, baby, drill" justifies the domestic exploration of oil from the areas that are designated as wildlife sanctuaries and off-limits to oil companies.

Given the theoretical points raised in this chapter, the fact that the lion's share of US domestic oil production comes from the least productive US oil fields, a relatively lesser quantity of output from the new explorations, such as the US Outer Continental Shelf and/or Alaska's National Wildlife Refuge (i.e., more productive US oil provinces) would neither change the center of gravity (the long-run price) nor markedly reduce the short-run price of oil in the United States. What happens is that the newly explored fields fetch differential rent based on the differential productivity of their oil relative to the center of gravity. And the long-run price would not budge a bit. Therefore, from the standpoint of environmental policy, we contend that drilling in the United States amounts to double trouble: (1) it would not change the long-run price of oil and (2) it would ruin the environment with incalculable social costs resulting from uncertain externalities, even when there is no catastrophic accident.

Finally, neither appealing to neo-Malthusian scarcity nor stretching the truth by invoking the tautology of "peak oil" in this elaborate interdependent system would hold any water. In this globalized world, neither an honest motivation for self-sufficiency nor the deceitful cry for power projection under the guise of national security (and strategic oil) leads to lower or higher oil prices and/or any significant provision of "secure" oil outside of the global system. Global oil is not kind to those who wish to have their cake and eat it too. To sum up, neither the fanciful drilling in Alaska's ANWR nor the fraudulent US invasion of Iraq would have a legitimate (objective) cause in getting access to oil for the sake of claimed self-sufficiency.

# 5

# Oil and Capital: "Logic" of History and "Logic" of Territory

*No social order is ever destroyed before all the productive forces for which it is sufficient have been destroyed, and new superior relations of production never replace older ones before the material conditions for their existence have matured within the framework of the old society.*

—Karl Marx
Preface to *A Contribution to the Critique of Political Economy* (1970: 21)

*Do international relations precede or follow (logically) fundamental social relations? There can be no doubt that they follow.*

—Antonio Gramsci
Selected from the *Prison Notebooks* (1971: 176)

## Introduction

In this chapter, the attempt shall be made to introduce and elucidate the epochal context in which the contemporary economy, polity and social relations fit. This is unequivocally essential for two reasons: (1) that the identity of the contemporary epoch brings to light the social and material transformations that are unique to our own time and thus to our own identity and (2) that the same formidable historical forces that tend to play a part as an arbiter of time also set the threshold on the dynamics of economy, polity, and power relations in our epoch. In what follows, the hint of "disjointed time" will postulate an analytical framework for a historical comparison of two separate phases of capitalism, namely, the age of imperialism—aided by the crutch of colonialism, and the plunder of "raw" and *unmediated* geography—and the epoch of post-Pax Americana/posthegemonic America, marching into the era of globalization.

This chapter offers a critique of Lenin's imperialism (and, its corollary, anti-imperialism) in light of its own epochal import and counters the subversion by overgeneralization of the concept that incorporates the past, present, and the future by self-styled contemporary Marxists. This not only misidentifies our present epoch but offers false claims about the period of history that is both unforeseen and untested with respect to what might be left of the shelf-life of capitalism. Here, the intent is to show that, from the standpoint of Marx's methodology, any generalization of imperialism beyond its own epoch would be a misrepresentation both of Marxism and of the epoch-bound Lenin's imperialism. This has a profound implication for the assessment of US (imperialist) behavior in the period since the fall of the Pax Americana (1945–79). This discussion is not simply for the sake of theory and intellectual exchange, but for the purpose of immediate political commitments that are bound up with the liberating movements of hundreds of millions of struggling masses that need to know the contour of global contradictions and the identity of their epoch.

## A Note on Methodology

A typical investigation begins with the observation of real (concrete) phenomenon as its point of departure. But a concrete, observable phenomenon is also composed of the unity of diverse and complex determination, which in reality is an outcome: a point of arrival. Therefore, if one were to theorize, one would have to abstract from the complexity of this concrete, "chaotic whole" to discover (in thought) the presupposed, simpler (abstract) categories that lay behind the façade of this ultimate determination. Yet, this abstraction will remain incomplete if one fails to reconstruct—via these simpler, abstract categories—this original observable phenomenon in thought: hence the double journey (i.e., a roundtrip) of moving from an observable concrete to the unobservable abstract and back to the observable concrete in thought.

Such an abstraction is not axiomatic (i.e., speculative); it is not approximate via the process of "successive approximation"; it is not a product of the mind's own ingenuity or ineptitude; it is precisely a *real* abstraction by virtue of being *mediated* through the appropriation of a *real*, concrete object of investigation by thought. As Marx put it, by entering into this (dialectical) journey, "the chaotic conception of a whole" would turn into "a rich totality of many [ordered] determinations and relations" (1973a: 100, 101–8; also Rosdolsky, 1977: 25–28, 561–70). It is well-known that Marx criticized Hegel for failing to show that the conceptualization of a real subject in thought does not cause its existence. Rather, the corporeal (concrete) subject itself is the very source of conceptualization, which then must be stripped of the encumbering concealments to be made into an abstraction, and thus a concept, which is a thought category within the grasp of the mind. This criticism also applies to methodologies that are embedded in logical positivism, methodological individualism, or idealism, particularly in mainstream (neoclassical) economics. Therefore, conceptualization need not be ideal, axiomatic, imaginary, or, for that matter, dependent upon a set of arbitrary and ad hoc assumptions. Within this framework, assumptions—and their possible role in theory—are all an internal corollary of the concepts themselves, not a figment of one's imagination. In others words, based upon this (materialist) methodology, a point of departure is the real subject itself in the anticipation of being observed by the perceiving mind, not the epitome of perceiving mind imposed on an independent reality in a speculative quest for discovery.

A pertinent issue in this chapter is to utilize the Marxian meaning of capitalist competition in the presence of the enduring concentration and centralization of capital in the oil industry—a view already dealt with in this volume. As we have noted earlier, the oil sector becomes further compounded by the presence of landed property and materialization oil rents in the process of valorization. Yet, the question of imperialism (the higher, not the highest, stage), is very much in conformity with our theory of periodization for the oil industry, from cartelization to competitive globalization. Here, our study of oil in its entirety confirms the validity of Marxian competition in advanced capitalism and thus the authenticity (and relevance) of the theory of value at the present stage of capitalism. Moreover, our theory shows the evolution of the industry from a cartel (an entity pertinent to imperialism) to the collapse of the cartel, and global competition in the presence of concentration and centralization of capital, which had undermined the subjectivity of a handful of capitalists in favor of the objectivity of the system as a whole. This finding is exactly the opposite of the "modification" of Marx attributed to Lenin, the origins of which are traceable to Hobson's unambiguous neoclassical orientation.

The oil industry was a mother of all cartels in the recent past. The observable present, being an outcome of this evolution, is bound to be entwined with the residues of past petroleum history. Hence, we need to reexamine the validity of our abstract categories that may have come to predate the present course of events and structure. Therefore, we must find an adequate measure of periodization for the production of oil, a measure that allows us to treat and investigate the present (i.e., the decartelized and globalized oil) as a distinct entity and, at the same time, as an evolutionary outcome of the past. The following passage sheds some light on the question of epochal abstraction and historical categorization in critical political economy:

> Bourgeois society is the most advanced and the most complex historical organization of production. The categories that express its relations, and an understanding of its structure, therefore, provide an insight into the structure and the relations of production of all formerly existing social formations the ruins and component elements of which were used in the creation of bourgeois society. Some of these unassimilated remains are still carried on within bourgeois society, others, however, which previously existed only in rudimentary form, have been further developed and [thus] attained their full significance, etc.... Bourgeois economy thus provides a key to the economy of antiquity, etc. But it is quite impossible [to gain insight] in the manner of those economists who obliterate all historical differences and who see in all social phenomena only bourgeois phenomena... In all [societal] forms in which landed property is the decisive factor, natural relations still predominate; in the forms in which the decisive factor is capital, social [and] historically evolved elements predominate. Rent cannot be understood without capital, but capital can be understood without rent. Capital is the economic power that dominates everything in bourgeois society... It would be inexpedient and wrong therefore to present the economic categories successively [i.e., in sequence of their historical presence] in the order in which they played a dominant role in history. (Marx, 1970: 210–13)[1]

To grasp the contemporary state of the capitalist mode of production, one must proceed from its presupposition, both in reality and in mind, to identify the *specific* categories that underlie its development. This allows the simpler categories to reflect both the complex and intensified relations of the developed concrete, as opposed to the undeveloped and slight relations of "immature" concrete. Money, for instance, made its presence prior to capital, wage labor, and modern landed property in historical time. Yet, it did not become a full-fledged category (i.e., an equivalent form) until the very development of capitalism (Marx 1970: 208). In other words, from the standpoint of methodology, one may not speak of money until one is able to develop the category of capital (as a social relation) theoretically; a category that is represented by money—an equivalent form. This very basic requirement, however, has been neglected by many, including Heinrich (2012: 63) who erroneously speaks of "monetary theory of value," instead of labor theory of value, and thus misrepresents Marx's view of value theory. Michael Heinrich simply violates the first principle of categorization in Marx's methodology and abandons the very question of historical materialism as to the assessment of capitalist mode of production. For an accurate treatment of *capital* and *money* in Marx's value theory at the introductory level see Fine and Saad-Filho (2010). For the debate on similar issues the reader is encouraged to see Kincaid (2007, 2008), and Fine and Saad-Filho (2008, 2009) as the rejoinder. Kincaid's slipup does perhaps parallel with Heinrich's in that he misperceives Marx's mode of presentation (of the categories) as an actual appearance of social relations that are subject to the material conception of history. In other words, *money* as capital (i.e., as an equivalent form) cannot be conceived other than as a figment of one's imagination, before *capital*, as a category and thus as a social relation. By the same token, it is the dominance of capitalist social relations that renders both the landed property a modern category and rent a valorized sui generis capitalist relation.

Methodology is a melded embodiment—just like pregnancy in which one cannot be pregnant and be expecting to a degree. Here, particularly on the subject of oil, it appears that unfortunately many within the heterodox economics traditions (including radicals, post-Keynesians, institutional economists, and neo-Marxists) are indeed impregnated by the orthodoxy. That is why, despite the critical question of oil as a subject, there has been neither a serious dialogue between orthodox and heterodox traditions nor a genuine discourse within the heterodoxy itself over the globalization of oil.

## REREADING LENIN'S IMPERIALISM

We wish to start with contemporary critics of globalization to differentiate between the embryonic and fully developed forms of "world market"—from a fetus, dependent and helpless, to an autonomous person unaided by the mother's body. This is a matter of qualitative change in the parlance of dialectics. Panitch and Gindin (2003) stage their critique of "globalization" by a citation from Marx as follows:

> This was famously captured in Marx's description in the *Communist Manifesto* of a future that stunningly matches our present: "The need of a constantly expanding market for its products chases the bourgeoisie over the whole surface of the globe. It must nestle everywhere, establish connections everywhere . . . it creates a world after its own image." But affirming Marx's prescience in this respect runs the risk of treating what is now called globalization as inevitable and irreversible. (2003: 4)

The authors then recite the interrupted history of the international system in this intervening time collapsed in the course of a revolution, two world wars and a Great Depression in between:

> The postwar reconstruction of the capitalist world order was a direct response on the part of the leading capitalist states to that earlier failure of globalization... During the brief postwar "golden age"... capitalist globalization was revived, and was further invigorated through the neoliberal response to the economic crisis of the 1970s. The outcome of this crisis showed that the international effects of structural crises in the accumulation are not predictable *a priori*. Of the three great structural crises of capitalism, the first (post-1870s) accelerated inter-imperialist rivalry and led World War One and Communist Revolution, while the second crisis (the Great Depression) actually reversed capitalism's internationalizing trajectory. Yet the crisis of the early 1970s was followed by a deepening, acceleration and extension of capitalist globalization... What this erratic trajectory from the nineteenth century to the twenty-first century suggests is that the process of globalization is neither inevitable... nor impossible to sustain... [W]e need to distinguish between the expansive tendency of capitalism and its actual history. (2003: 4–5)

Since this particular position is tangent to both the geographical expansion and historical evolution of capital, it would be necessary to inspect it more closely. First, there is a paramount methodological question lingering in the first citation concerning the "constantly expanding market" and the chase "over the whole surface of the globe." This citation is often taken as globalization of the social relation of capital in that early embryonic stage, which is both historically and ontologically untrue. Those who are familiar with Marx's method should know that he is talking about the potential that will eventually turn to actuality by concrete social and material transformation. At the stage that Marx is alluding to this global pursuit merely one form of capital was beyond the border of the nation-state—commodity capital. This is the Ricardian era that has nothing to do with the globalization of today. Marx here is speaking of embedded universality of capital as a social relation; markets (including world market) are the medium and mode of expression through exchange.

Second, beyond the popular pamphlet mentioned above, the method employed in Marx's critique of capitalism is synthetic, not axiomatic. Therefore, actual history is the demarcation for Marxian categories to reveal themselves. An example is the internationalization of yet another form (and circuit) of capital, namely, finance capital, that emerges despite the massive disruption by the First World War. This also carries with it the role of the state in the accumulation process. And notwithstanding the Great Depression and another massive disruption, namely, the Second World War, the internationalization of a productive form (circuit) of capital puts the transnationalization of social capital on the front burner. It is imperative to realize that for capital (a social relation beyond boundaries of the nation-states) to become a de facto global entity, it is essential that there will be a global social circuit in place in terms of *commodity*, *money*, and *productive forms* so that unification of the spheres of circulation and production at large will become possible. These forms (of capital) are but manifestation of the movement and thus moments of social capital as a whole in *transformation*. (Palloix 1977; Bina and Davis 1996, 2008).

Third, on the question of theory and methodology, let us defer to John Barrow, the author of *Theories of Everything* (1991: 272) and one of the brightest astrophysicists of our time, who said: "The physical theories that we employ to understand the Universe are always synthetic. They tell us things that can only be checked by looking at the world. They are not logically necessary." We expect that Marx's value theory and our theory of globalization follow the same path. Finally, from our perspective, the notion of predictability raised by Panitch and Gindin (2003) is not equivalent to a teleological reading of history. From a materialist approach to political economy,... theory always takes a backseat as the note on methodology at the opening this chapter shows. We cannot predict what we cannot know due to contingencies. What we know are the potentialities that are embedded in economy, polity, and social relations and that they could very well turn to actuality. And there is not much more one can ask of a scientific theory either in physical and biological sciences or in social and cognitive sciences.

We have shown through the medium of oil (indeed of petroleum and energy) in its stage-by-stage evolution (i.e., through economic, political, and ideological transformation) how and when shifts of power (and power relations) have taken shape for nearly a century. These shifts manifest the progression from the division of world oil among monopolies (and national cartels), as Lenin rightly identified in his *Imperialism* (1970 [1916]), to the value-theoretic world of oil—decartelized, competitive, and free of hegemony or imperialism. But V. I. Lenin was very busy with severe and multifaceted political debates on imperialism, and with extraordinary revolutionary upsurge and his own revolutionary fervor that in the spare of the moment praised Marx (Marx's theory of competition) for a very wrong reason. We read:

> Half a century ago, when Marx was writing *Capital*, free competition appeared to the overwhelming majority of economists to be a "natural law". Official science tried, by a conspiracy of silence, to kill the works of Marx, who by a theoretical and historical analysis of capitalism had proved that free competition gives rise to the concentration of production, which, in turn, at a certain stage of development, *leads to monopoly*. Today, monopoly has become a fact. (Lenin 1970 [1916]: 17–18, emphasis added)

This is not the only place that Lenin speaks of "free competition"—a notion that had newly been adapted by the early neoclassical economists, such as Hobson (1891) and Clark (1891) and which found contrast and a point of departure for hitherto the monopolistic models in mainstream economic theory. From the standpoint of methodology, Lenin's references, particularly on Hobson (1961 [1902]), to observed evidence on monopoly and cartel are not only limited to the empirical aspect of the issues at hand, but also speak quietly but surely on how little Lenin understood Marx's theory of competition and how prematurely the "decay" of capitalism may relate to a stage where nearly four-fifths of world's population had neither lived under capitalism nor had a foggiest idea what capitalism would look like. Therefore, it is imperative, as Panitch and Gindin (2003: 5) stress, "[to avoid writing] theory in the present tense." And on this very issue, Karl Marx offers an exceptional model of theorizing, with respect to the *tendencies* and *potentialities* of capitalism, beyond his own temporal experience and time in his magnum opus, *Capital*.

Nevertheless, our contention is that it is awfully difficult, if not downright impossible, when one's epoch is seen as "monopoly," with formidable cartels running roughshod over economy and polity, not to be tempted or captivated by the (transitory) concrete; and also it is worth bearing in mind that in this transitional stage (of yet undeveloped capitalist competition) someone of Lenin's stature and aspiration, or that of Rosa Luxemburg's, must have felt an urgency to turn the rotten colonial/imperial system on its head once and for all. In the very same context, it is not at all surprising that Luxemburg, one of the sharpest revolutionaries of all time, in the spark of this revolutionary era, should formulate a theory of capitalist breakdown (1951 [1913]) that does not square with Marx's theoretical framework. This teaches us that, while we should recognize these incomparable revolutionaries, we also should take the responsibility of assessing their words critically and in the context of the time in which they had been put to print.

Therefore, aside from the slipup on Marx's (synthetic) competition—which is not the same as "free competition" and does not constitute as a point of departure for Hobson's monopoly—Lenin is right on the mark on imperialism and cartelization of key industries. As we have shown in the previous chapters of the book, cartels are entities in which the pricing and other pertinent matters related to profit making are fixed and preordained by the executive committees and syndicates. These entities are the market onto themselves. For instance, the International Petroleum Cartel was essentially the world market from its inception, in 1928, until its collapse in 1972. Therefore, in such a circumstance there is no room for a theory value (in a Marxian fashion) to make a determination. The discretion of the syndicate through administered prices is a stand-in for the forces that are yet to be and yet to manifest the materialization of a theory of value in operation. Here Marx was silent and Lenin did the talk—so to speak. That is why, far from being the highest stage of capitalism, the division of the world among cartelized monopolies is an epoch limited to imperialism. Conversely, the epoch of imperialism is a specific era in which cartels are in charge. In this era, competition had not been developed adequately so that capital could turn into a social relation in a widespread and universal fashion. As a consequence, the "free competition" attributed to the period of 1860–70 by Lenin is not based on Marx's theory of competition in capitalism. Yet Lenin observes:

> [T]he principal stages in the history of monopolies are the following: (1) 1860–70, the highest stage, the apex of development of free competition; monopoly is in the barely discernible, embryonic stage. (2) After the crisis of 1873, a lengthy period of development of cartels; but they are still the exception. They are not yet durable. They are still a transitory phenomenon. (3) The boom at the end of the nineteenth century and the crisis of 1900–03. Cartels become one of the foundations of the whole of economic life. Capitalism has been transformed into imperialism. (1970 [1916]: 20)

Lenin also juxtaposes "competition" and "monopoly" side by side and utilizes them rather effectively against the soft and smooth concept of "ultra-imperialism" by Kautsky. Yet, this ingenious political move comes at a not-so-ingenious theoretical cost. It misses the synthetic meaning of competition. Lenin writes: "[T]he monopolies, which have grown out of free competition, do not eliminate the latter, but exist above it and alongside it, and thereby give rise to a number of very acute, intense antagonisms, frictions and conflicts" (105). Despite its value in class analysis, the same *juxtaposing* problem has also been reproduced in Poulantzas (1975, Part II, Ch. 2), where the author muddles real competition in capitalism by invoking "competitive capitalism" and "monopoly capitalism" side by side (in an absurd dichotomy), and by overuse, and sometimes utter abuse, of "mode of production" versus "social formation." And this was not a solution but probably an intellectual distraction that detained us from the possibility of discovering the eye of the storm that was about to converge on the old Pax Americana era—the storm that strikingly removed and reassembled us at the onset of a new epoch to which the present was about to enter in the mid-1970s.

Another germane issue in Lenin's *Imperialism* is the idea that monopoly acts as a transition to a "higher order" beyond capitalism and that somehow this would be the end of capitalism. This would have far-reaching implications for Lenin's vision, then, concerning the future society and economy in the postrevolutionary era. And although Lenin used other phrases of lesser finality for imperialism, such as "the latest stage of capitalism," in light of the following passage, the subtitle: "the highest stage of capitalism" is more fitting. We read:

> We have seen that in its economic essence imperialism is monopoly capitalism. This in itself determines its place in history, for monopoly that grows out of the soil of free competition, and precisely out of free competition, is the *transition from the capitalist system to a higher socio-economic order*. (Lenin 1970 [1916]: 148 emphasis added)

Another salient question in imperialism, as the "highest stage of capitalism," is the existence of differential rates of profit between monopolies (and cartels) on the one hand, and nonmonopolies on the other. To solve this problem, one has to untangle the question of the "epoch" of imperialism from the alleged "highest stage," thus clarify the validity of Lenin's epoch as imperialism while demonstrating as to whether it was the highest stage of capitalism; and if so, then, does the present epoch (as claimed by the many self-styled Marxists) also qualify for the same connotation and category. There is no question that one of the pillars in Lenin's *Imperialism* is "monopoly" profit (for investigation of differential rate of profit see Moudud, Bina, and Mason, 2013). In other words, without monopoly profit there is little recourse to identify the monopoly of a cartel. As has been demonstrated in detail in this book, formation, actions, and aspirations of the International Petroleum Cartel (IPC) (1928–72) perfectly dovetail with Lenin's account of cartelization and also confirm that this "monopoly [had] grown out of colonial policy (Lenin 1970 [1916]:149)," as the latter indicated. Indeed, Lenin would have been elated had he seen our analysis of old oil concessions under Achnacarry. In a nutshell, the story of the IPC is a forceful corroboration of a portion of Lenin's thesis.

This implies that imperialism, as Lenin rather fittingly articulates, is a "special" stage in the development of capitalism—not a "policy" in Kautsky's rendition. But, contrary to Lenin's thesis, at this "special" stage, which is built into the structure of capital, competition had not yet been developed adequately and thus it would be mistaken to call it the highest stage of capitalism. In other words, it may be a higher state but not the highest for two pertinent reasons: (1) that this is not informed of Marx's theory of competition despite Lenin's repeated claims to the contrary and (2) that any *materialist* methodology worthy of the name could not submit to *speculation* about the distant future (Marx 1973b [1847]: 128–34, Marx 1973a [1857–8]: 100–8, Bina 2013, Weeks 2011, and Weeks 2013). Lenin misunderstood Marx when he declared: "Free competition is the basic feature of capitalism... *monopoly is the exact opposite of free competition*" (Lenin 1970 [1916]: 104–5, emphasis added). Marx's competition is synthetic and has nothing to do with the opposing poles of "competition" and "monopoly," which were conceived by the early neoclassical economists and contaminated Lenin's work, perhaps through Hobson's or even Hidferding's work (see Bina 2013 on Marx's synthetic competition and axiomatic pole of competition/monopoly in neoclassical economics). As a consequence, imperialism is certainly neither a "policy" nor a stage of "decaying" capitalism—when an overwhelming majority of humanity had not yet experienced it. Far from being monopoly, today's global capitalism is substantially higher (as a stage) than capitalism in Lenin's era while, at the same time, considerably more concentrated and more competitive than the latter (see Moudud, Bina, and Mason 2013).

Finally, the most devastating influence of Lenin's *Imperialism* is the advent of what has subsequently become known as the *stagnationist thesis*—a tendency to stagnation (and decay) in mature capitalism. This, shall we say, subthesis has made a great deal of splash in economics literature beyond Lenin's outline. We read:

> As we have seen, the deepest economic foundation of imperialism is monopoly. This is capitalist monopoly, i.e., monopoly which has grown out of capitalism and which exists in the general environment

> of capitalism, commodity production and competition, in permanent and insoluble contradiction to this general environment. Nevertheless, like all monopoly, it inevitably engenders a *tendency of stagnation and decay*. Since monopoly prices are established, even temporarily, the motive cause of technical and, consequently, of all other *progress disappears* to a certain extent and, further, the economic possibility arises of deliberately *retarding technical progress*. (Lenin 1970 [1916]: 119, emphasis added)

This influence has a Malthusian lineage (with overtones of underconsumption), which appears to have derived from Hobson (1961 [1902]) or even Hilferding (1981 [1910]), and which unwittingly tainted the works of Luxemburg (1951 [1913]), Keynes (1964 [1936]), Sweezy (1942), Steindl (1952), Kalecki (1954), and Baran and Sweezy (1966) among others (see Bleaney 1976). And it is hardly an overstatement to say that almost all stagnationists on the left are either Keynesian or Leninist today. Zoninsein (1990) presents an articulate assessment of Hilferding's view of "monopoly capital."

## IMPERIAL FETISH: LEAPING OUT OF TIME

Imperialism, hegemony, the Pax Americana, and the epoch of globalization are all historical concepts in need of validation by a concrete and specific period of living history; hence, each of these concepts is historically constrained (or, conversely, inspired) by the concrete reality itself. Imperialism, particularly of Leninist lineage, is neither timeless nor malleable; therefore it cannot be arbitrarily assigned to any epoch in the latter stages of capitalism.[2] Imperialism—as a system, as opposed to a policy (thanks to Lenin)—essentially belongs to a historical period in which the world is divided among imperial powers and, via their national syndicates, consortia, or monopolies, extend their contentious imperial interests. This particular period had already prompted its contradictory effects in terms of the transnationalization of social circuit of capital in all three forms, namely, commodity capital, finance capital, and productive capital (Palloix 1975; Bina and Yaghmaian 1988, 1991).[3] Hence, even accidental emphasis on the sui generis export of capital would take us back to the period in which the scope of capital, as a social relation, was merely limited to the national boundaries. Hence, speaking of "export" of capital makes sense in the period of imperialism, as acknowledged by Lenin and others. For a critical survey of traditional theories of imperialism see Brewer (1980).[4]

Today, in contrast, the transnationalization of social circuit of capital is an accomplished fact and hence it would be ludicrous to rely on a tiny subset of such relations, that is, "export of capital" as a distinguishing feature of our time. Today, neither "Americanization" nor American hegemony or imperialism, as defined by Lenin or Bukharin,[5] has material, social, or ideological wherewithal to withstand the social relations of capital across the world. The vessel of Pax Americana—that is, the grand container of economy and economics, polity and politics, and the international relations of the postwar era—has been overpowered by social relation of the contained. The tree has outgrown its pot. We have now one foot in the past and another in the future, and the present is disjointed much like warped space. And an elementary view of this is to traverse between the "doctrine of containment" and the disorder of "preemption," irrationality, and predatory international politics across the global polity. If one calls this an *epoch* of imperialism, we may rank anyone who turns against a bully on the block an anti-imperialist. To be sure, all these actions may well put the bygone imperialists to shame by scope of their audacity and scale of sheer violence. But these actions are devoid of an epoch of their own in our time. That is why they are already doomed at the inception long before being carried out. This is not worthy of any self-respected imperialist whose legitimacy is simply dependent upon the fact that it ought to live in one's own epoch.

The kernel of Lenin's *Imperialism* is an outright and widespread colonization and division of the world among the "great powers." This is in tandem with outright and widespread cartelization and monopoly, spreading across national boundaries, and attachment to colonizing nation-states and their foreign policy.[6] This is in stark contrast with today's global capitalism in which capital has already been transnationalized beyond monopoly, cartel-like composition, and the assorted direct (i.e., unmediated) administered frameworks. This, for instance, can be seen from the collapse of the most notorious cartel in recent history, namely, the IPC, just before globalization of the oil industry in the early 1970s. In the epoch of "ascendancy" of finance and national cartels, outlined by Lenin, the reliance on colonization and unmediated control shows that the *law of value* has not yet become operational in nearly four-fifths of the world.[7] This implies that capital as a social relation has yet to be transnationalized for the law of value (and capitalist competition) to take hold universally across the globe. This epochal departure is also commensurate with worldwide class polarization and transnationalization of labor power across the global landscape. Therefore, it would be a mistake to draw a parallel between

Lenin's epoch of imperialism, via finance, and the 2008 financial crisis.[8] As for "free competition, as opposed to monopoly," Lenin's own perception, notwithstanding the larger intellectual climate of the time, which appears to some extent parallel with liberal economists of his time, like Hobson, as often tended to idealize the period of 1820s through 1860s in England as "free competition." This is contrary to Marx's competition according to which capitalist competition had not yet sufficiently developed in this period. This is precisely what mainstream economists of today label as competition, which is in fact a benign state of affairs unlike real competition in capitalism. This is how "competitive capitalism" and "monopoly capitalism" find their dichotomy in this subtle mistake. And how Marxian economics has split along this very methodological line on the meaning of competition and monopoly ever since. For Marx, however, capitalist competition gains strength as capitalism develops further toward maturity with concentration and centralization of capital—not in its infancy, where there is little muscle flexing among capitals. And, despite a brilliant depiction of imperialism—as a mere prehistory of today's global capitalism—Lenin's tour de force has no affinity with Marx's conception of capitalist competition—a synthesized manifestation of concentration and centralization of capital in reciprocal dissention and discord. In other words, for Marx, war of capital upon capital gains more intensity through the imperialist stage and continues as competition advances further in mature capitalism.

Does this mean the post-imperialist world would be free of war among contending powers in the newly emerged polity? By no means—this only reveals that the potential driving force and underlying antagonism behind the incessant diffusion of power tend to transform the power structure by enhancing its capacity to resolve unnecessary disruptions, and by internalizing conflicts through mediation deep within the polity, which ipso facto bring to light the futility of naked aggression and preemption. This is not unlike the motion of large and seemingly autonomous heavenly bodies, in quantum cosmology, that are subject to all possible quantum changes in the entire universe. This surely creates more uncertainty, yet there are also more possibilities. Therefore, speaking of imperialism as an inanimate object—that is, in frozen figure of past social relations—characterizing our epoch of globalization of social relation of capital smacks of (imperial) *fetishism*.[9]

We have to be reminded of Marx's own expression, "the conquest of mode of production," a universal and omnipotent goal of capitalism. With Marx's theoretical insights in mind, the raison d'être of capitalism is not about the *means* but the *ends*, as the means are always obtainable by capital's own transformation.[10] That is why an overcoming of (external) barriers to capital accumulation has always been pursued through transforming the "means," such as the ability to do away with outright colonialism. Therefore, this historical objective had to take precedence over direct territorial (colonial) conquest, just as newly established capitalism in the midst of nineteenth-century America took precedence over anachronistic slavery in freshly established American plantation systems both prior and under the secessionist confederate states in the South. Likewise, contemplating a "need" for *direct physical control and access* to Middle East oil (or other raw materials at large) in this day and age is not unlike window-shopping for the strongest and brightest of (human) slaves in the hustle of Christmas, say, in the capital of the Confederacy, West Broad Street, Richmond, Virginia. This should say enough about the absurdity of this proposition. Yet such empty talks, that is, discourses that gyrate about the fetish of "access," come from respected quarters that are not the least suspected of irrationality and lack of common sense. These conversations are empty of truth, because they are neither informed nor observant of valorization of the landed property (of oil deposits) under one universal rule—capital permeated as a social relation. They neither appreciate the power of plasticity nor grasp awesome germination and proliferation of capitalism that makes the control of oil, in whatever shape and form, next to impossible. As has been demonstrated earlier in this book, whatever happens within the synthesis of capital and landed property in oil (i.e., in the process of valorization), it would have little to do with such vulgar explanations. As a consequence, today, virtually nothing in the global polity can be exempt from this universal rule—if capitalism as a modern social relation were to mature and to grow out of its imperialist past and colonial infancy.[11]

At the same time, tendency for "inter-imperialist rivalry," which was apparently a focal point of contention in the early part of twentieth century and debated vigorously between Lenin and Kautsky, had already encountered its countertendencies and thus transmuted through steady global interdependence, particularly through an intense and irreversible transnationalization of capital. This should have implications for the notion of "bloc formation," as one of many historical forms, which itself is a derivative state and stage of capitalist competition in conjunction with development of capitalism on a global scale. In any case, in the longer run it is a competition deep within social relations (of capital) that trumps all forms and moments of rivalry (if any) in global polity. Hence the reader should note that references to competition in this book are not equivalent to "rivalry," observed by Lenin and Bukharin, among others. As for colonial and semicolonial powers in Lenin's *Imperialism*, it boils down to epochal validation as opposed to epochal invalidation of such powers, and as was demonstrated in this book, such invalidation has been horrendous for the United States. The epochal change has no magic wand for peaceful disposal of a great power that is already invalidated and has to stand down; and, if history is of any consolation, the arbiter of time and disappearance of façade will ultimately do the trick. This is not in the slightest an easy task for emerging global polity or for world peace and stability in the near future.[12]

## ORTHODOXY AND THE EUTHANASIA OF CAPITALIST COMPETITION

It is no exaggeration to declare that competition is among a handful of concepts that ravaged both orthodox and heterodox economics. Within the orthodoxy, "perfect competition," while it obtained the status of an axiomatic model (logical and timeless), nevertheless stood against the evolution and particularity of capitalism. An axiomatic spectrum of competition/monopoly became a standard model for the calibration of real capitalist enterprises that are placed nearer or further from these two imaginary poles. This method (with respect to calibration of competition) holds an uncanny resemblance to the Procrustean bed—an iron bed used by Procrustes ("The Stretcher") for modification of the human body consistent with the size of a chosen benchmark. Yet, at least in this mythological story, Procrustes used a real bed as a measure, not an imaginary one.

The above parable is useful for identification of orthodoxy, and some of its counterparts within heterodoxy, from the standpoint of dealing with capitalist competition. The spectrum of competition/monopoly is said to be a heuristic device for calibration of real competition in capitalism. Yet, departure from the (illusory) "perfect competition" is a demarcation that also grants concrete identity to the departed, that is, "imperfectly" competitive firms—an identity that is other than competition.[13] In other words, the door had already been closed to the possibility of competition in capitalism by such an embrace. And no amount of departure from "perfect competition"—displayed through "appreciation" of realism—ever disentangles the subject from the grip of orthodoxy. Consequently, our criticism does not end with neoclassical economics, a bastion of orthodoxy, but extends widely to all self-styled heterodox economists who survive on this or any other stratagem in denial of competition in mature capitalism.[14]

At this juncture, let us shift gears and take up the question of competition from Marx's standpoint, in the presence of concentration and centralization of capital.[15] The emphasis on presence of the latter renders competition genuine and germane to the accumulation of capital. Thus, as capital widens and deepens further, first within a single production unit, then within a multiplant enterprise, and in due time beyond an industry, a collection of industries, a national economy, a region composed of many nations, and finally across the world, competition gains overwhelming momentum, and breathtaking intensity, like the wild fire does in every direction in a dried up forest. According to Marx, competition is not passive, it is not a bundle of harmony; competition is not a state of affairs but a contentious and contending process arising from inner mutuality of capital in aggregate, which manifests by individual capital's interface and interaction. In other words, capitalist competition is a synthetic enterprise. Aside from the genesis of Marx's observation in *The Poverty of Philosophy* (published in 1847), a treatise on method and vision about capitalism and a reply to Proudhon's *The Philosophy of Poverty*, it is imperative to note that competition in Marx is the process (and inner force) by which capitalism renovates and restores itself toward maturity. Therefore, the maturity of capitalism, particularly in its global stage, is but an insignia of competition, not monopoly. But the indoctrinating vision of neoclassical competition, reinforced by the anachronism and misapplication of Lenin's *Imperialism*, created a fertile ground for this false and contrary impression across the board in nearly all social sciences, including economics. For Marx, capitalist competition is historically advancing, not retreating. How historically? The following passage is instructive:

> What competition brings about, first of all in one sphere, is the establishment of a uniform market value and market price out of the various individual values of commodities. But it is only the competition of capitals in different spheres that brings forth the production price that equalizes the rates of profit between those spheres. The latter process requires a higher development of the capitalist mode of production than the former. (*Capital*, vol. 3, 1981: 281)[16]

In the recent issue of *The Socialist Register*, David Harvey writes:

> When demand and supply are in equilibrium, he [i.e., Marx] argued, they cease to explain anything while the coercive law of competition functions as the enforcer rather than the determinant of the general laws of motion of capital. This immediately provokes the thought of

> what happens when the enforcement mechanism is lacking, as happens under conditions of monopolization, and what happens when we include spatial competition in our thinking, which is, as has long been known, always a form of monopolistic competition (as in the case of inter-urban competition). (2011: 7)[17]

Let us identify a number of problems in this tiny passage. First is the question of market demand and supply and their accidental equilibrium in a disaggregate economy such as the capitalist one. Even central planning cannot run a tight ship as to equilibrate demand and supply of the commodities on a daily basis. And this is crystal clear for Marx who describes: "If demand and supply cancel one another out, they cease to explain anything, have no effect on market value and leave us completely in the dark as to why this market value is expressed in precisely such a sum of money and no other" (*Capital*, vol. 3, 1981: 291). The context of this statement is the contention that relying on equalization of demand and supply—the fact that more and more economists have lived in this self-delusion ever since—is no more than a tautology and that the "inner laws of capitalist production clearly cannot be explained in terms of the intersection of demand and supply" but by the value (or, in more developed capitalism, prices of production) as the center of gravity of fluctuating market prices. Harvey does not appear to grasp this elementary point that short-run market prices are only accidentally equal to the magnitude of value in commodities.

Second, Harvey appears confused on the role of competition between "enforcer" and "determinant" with respect to "the general laws of notion of capital." Marx is very explicit about what capitalist competition does or does not do. He observes: "Competition executes the inner laws of capital; makes them into laws toward the individual capital, but does not invent them. It realizes them" (*Grundrisse*, 1973a: 752). Marx maintains consistently: "Competition generally, this essential locomotive force of the bourgeois economy, does not establish its law, but is rather their executor. Unlimited competition is therefore not the presupposition for the truth of the economic laws, but rather the consequence—the form of appearance in which their necessity realizes itself...Competition therefore does not *explain* these laws; rather lets them be *seen*." (Ibid. 552, emphasis in original). Again, he declares: "Conceptually, *competition* is nothing other than the inner *nature of capital*...the inner tendency as external necessity" (Ibid. 414, emphasis in original). Marx continues: "[T]his much is clear: a scientific analysis of competition is possible only if we can grasp the inner nature of capital, just as the apparent motion of the heavenly bodies are intelligible only to someone who is acquainted with their real motion, which are not perceptible to the senses" (*Capital*, vol. 1, 1977: 433).

Third, there is little chance of misunderstanding the above statements if one appreciates the process of value formation in mature capitalism. Such appreciation, which is contingent upon a meaningful critique of neoclassical theory, is sometimes lost on the left, particularly the radical left. Therefore, Harvey's confusion in the above passage is not an exception. He focuses exclusively on the meaning of "executer" as "enforcer" and with an anticipation that a particular state of affairs, namely, capitalist competition is in need of execution. Then, Harvey's deduction tells him that the absence of such enforcement is because of "conditions of monopolization," and by blending "spatial competition" in the mix, as a radical geographer, he reaches a conclusion that has been known to him—"monopolistic competition." That is how Harvey throws to his impressionable audience swirls of neoclassical economics fakery in the name of Marxism. As Ben Fine readily documented, "Harvey's contradictions are not always as dialectical as they might be. For internal contradictions, at the economic level, Harvey presents the three volumes of *Capital* as successive models rather than as the reproduction in thought of simple categories at more concrete and complex levels . . . This means that the unity of his treatment is flawed." (Fine 2006: 138). The implication of such methodology is probably not more damaging than in Harvey's *The New Imperialism* (see also Arrighi 1994, a volume that essentially falls within the same category).[18]

As can be seen, Harvey has shown little appreciation for the subject of capital accumulation, which is tied to the dynamics of the inner nature of capital (in aggregate) captured by Marx's value theory and displayed through competition in capitalism proper. And while, for instance in *Monopoly Capital* (1966), Baran and Sweezy, as pupils put an end to Marxian analysis, set forth by the master himself, in a swift and painless strike, Harvey has taken his time by several decades of on-the-one-hand and on-the-other-hand hesitation before artfully entombing the value theory for good.

Now, let us shift gears and focus on Milios and Sotiropoulos's *Rethinking Imperialism* (2009) in which competition in Marx's sense is appreciated by critiquing Monopoly Capitalism in many of its trappings and manifestations, yet remains unsatisfactory from the standpoint of historical materialism. This volume speaks of "historical transformation of power balance," yet it fails to recognize that the theory of value is a materialist theory, which comes to be further realized with the development of capitalism. Hence, it succumbs to Althusserian idealism in contradistinction with Marx's real abstraction and concretized theory. Milios and Sotiropoulos inform us that the transformation in capitalist mode of production (CMP) has already happened from production of the absolute surplus value to that of the relative surplus value; all we have to be worry about is to attend to the question of balance of power. They write:

> The transformations we have described...distinguish the form of capitalist domination even in the first period after industrial revolution in the nineteenth century (capitalism of absolute surplus-value) from the later form of this domination (capitalism of relative surplus-value). That which was transformed is not the "laws" of capital accumulation corresponding to the CMP, or in other words the structural characteristics of capitalist relations at all social levels, *but the conditions and forms of appearance of capitalist relations in the historical perspective.* In other words it is *a question of historical transformation of power balance* and accordingly of the organizational forms of power in developed capitalist social formations. (emphasis added, p. 129)

This passage is indeed a smoking gun, that is, a quintessential admission of Althusserian propensity to idealism advocated by the authors. Here is why: (1) by invoking the form of "domination," the authors insinuate authenticity and obfuscate the centrality of the overcoming by capital of all its "natural" limits (including the limitation of "working day") in concrete historical reality; (2) by ignoring the limitation of the working day, they miss the opportunity to see the battle over cheapening of labor power, the essential meaning of technological change in CMP proper—this would obscure the real meaning of intra- and intersectoral competition at the level of theory and across the global divide; (3) by introducing "a theory of modification of competition" (Ch. 8), the authors do not realize that all *modifications* are but the very weapons of capitalist competition; (4) by invoking the notion of "imperialist chain" (an idealized generality), they mock both the period that is characterized as an epoch of *imperialism* (e.g., by Lenin and others) and its aftermath, and turn a blind eye on the significance of the transnationalization of capital and contemporary globalization; (5) by idealizing Marx's theory of value, they ignore concretization at the level of the theory (both with respect to accumulation and class polarization), and neglectfully relegate nearly all attributes of transforming capitalism to "social formation," and then dismissively label them as "historicism."

To be sure, *Rethinking Imperialism* is very much justified in critiquing Monopoly Capitalism theory, center/periphery thesis, "new imperialism," and Imperialism, as the "latest" stage of capitalism. The book attempts to demonstrate that competition is a necessity in capitalism, despite the denial of a "stage theory," and the alleged inapplicability of a theory of periodization by the authors. Milios and Sotiropoulos also resolutely point to the shortcomings in Harvey's (and Callinicos's) notion of "geopolitical competition" and its implication for today's globalized world. But unfortunately, they misunderstand Schumpeter's theory of innovation via "creative destruction" (for instance, by lumping it together with Hilferding's)—and Schumpeter's theory of competition—by unnecessary cherry-picking and focusing on his theory of imperialism alone (see pp. 62–69, pp. 70–77, p. 143). *Rethinking Imperialism* still is antithetical to the many monopoly theories that come to contrast with the notion of competition in contemporary capitalism. Yet, it succumbs to an ideal type, not quite unlike the neoclassical framework, by degrading the *synthesized* and lively notion of competition in Marx to its imitation in Althusserian (idealist) terms (see Marx, 1969, 126–34; on method see Marx, 1973a, 100–8). In a way, "successive approximation"—the method of mimicking the ideal by the concrete—is remarkably at work in this book due to shaky foundation of Althusserian Marxism (see Benton 1984). Hence, a good deal of justified criticism against many monopoly approaches in radical political economy has bumped into the self-inflicted contradiction and methodological dead-end. A fascinating volume of heartfelt polemics, E. P. Thompson (1978) may provide the reader a not-so-disinterested glimpse of Althusser's lifelong ideological tendencies and political standpoints.

Consequently, both *The New Imperialism*, by Harvey, and *Rethinking Imperialism*, by Milios and Sotiropoulos, are out of sync and without the social and material underpinnings of capitalist competition and the significance of imperialism (and monopoly) as a *specific* epoch of capitalism. The first reinvents a "new imperialism" after imperialism's old image; the second trivializes imperialism by the denial of its import as a specific epoch in the history of capitalism. Figuratively, Harvey falls off the roof by walking to the front; Milios and Sotiropoulos, on the other hand, fall off the roof by walking backward to the rear.

## SECURITY OF OIL OR SECURITY FOR CONTROL?

*Security* and *oil* are one of inseparable pairs of words that are often discussed and written about in conventional international relations literature. But the *ideological* ramification of such inseparability has rarely been the subject of scrutiny in both economics and political science literature. On first blush, maintaining security appears an imperative matter for production, exchange, and transportation of any commodity across the global divide. Yet, nearly all quarrels over security, particularly in political science literature, have been markedly rotated about security of oil, to be more precise, security of oil from the Middle East or North Africa. Aside from the fact that a majority of scholars who write about security of oil show not much proficiency in particulars as to where yesterday's oil (and oil industry) was, as opposed to where today's oil is, it is curious why the excruciating cry for oil security should come from such quarters, especially in the United States. Indeed, cry for security of oil is now a cottage industry with ample upkeep from public and private expenditures in academic research, political fronts, think-tanks, lobby groups, and the US government. The concern for security is also widespread among a few governments that are known as the so-called allies of the United States. In this context we read:

> If the *International Energy Treaty* was the foundation for the development of a global energy security system, then the development of the producer-consumer dialogue represented the next stage in its development. (Yergin 2011: 273, emphasis added)

To know why oil has been singled out in the ongoing drama of today's security, one has to travel backward in time to be in the same room with the same gentlemen who broke ground for the IPC on the morning of September 16, 1928, in the Achnacarry Castle, Scotland. This meeting was of course secret, indeed top secret. But it was not secret to all pre- and postwar US administrations that cherished it as a godsend and treated it as a Trojan horse from heaven for conducting a uniform foreign policy, based on what is now dubbed by the belated cheerleaders (see Yergin 1991, 2011), as the Postwar Petroleum Order. There are unimpeachable historical accounts (and records) of this meeting (and more meetings and countless "memoranda of principles") that led to the outright control of world oil in an intricate, interlocking, and an aggregate international cartel (see Blair 1976, Engler 1977, O'Conner 1955, Tanzer 1969). Control, therefore, was the very mechanism by which the so-called Western world obtained total security of oil. And since the Achnacarry and up to 1972, the time that the IPC was no more, a universal vehicle for expedient US foreign policy through oil (security), had duly been in place; following the US ascendance to the summit of hegemony in 1945, the United States—despite nagging disapprovals by US Justice Department—had kept a cozy relationship with the IPC even prior, but certainly, and especially, after the Second World War. This relationship had been further solidified into an umbilical cord that tied cartelized oil to an indubitable US foreign policy in the Middle East and elsewhere in the oil producing regions.

The umbilical-cord analogy cannot be more fitting than in the case of nationalization of oil in Iran, in 1951, and the 1953 CIA coup d'état as a retort. This is a crystal-clear case as to the cause and effects of the US coup, and given nearly all US and UK declassified government documents and testimonies from witnesses and other pertinent sources of information on Iran and elsewhere, oil (i.e., IPC) had been the very cause of the US coup d'état—not the unsupported Cold War threat. The latter was employed as a parody, first by the British and subsequently by the Eisenhower administration, and then exploited as an ideological crutch and facilitator to get the job done. The question of Iranian oil nationalization then was that Iran wished to produce and sell its oil freely (based on competitive bidding, etc.) in the international market, away from neocolonial grip of the Anglo-Iranian Oil Company (AIOC)—a conjoined slice of the IPC. In other words, Iranians wanted to get rid of this bizarre colonial control and to abide by competitive market forces, which were nascent and percolating by the soaring entrance of the "independent oil" companies at the time. In essence, this was Mossadegh's dilemma as it was Gandhi's before him. But the Eisenhower administration wanted this *control* so desperately that it had been willing to overthrow the Mossadegh government in a clandestine operation and kill the germinating roots of democracy in Iran for an unforeseeable future. This is precisely how security of oil stands-in for cartelized control in the unconscious intellect of the disinterested public. But this is half the story; the relic of neocolonial past as to the control of oil still reiterates and echoes in nostalgic bursts in the concerns and conduct of US government in our postcolonial polity and posthegemonic America. This reverberation is the second half of the story of "security of oil" today.

Again the priority for safeguarding the principles of "As Is" agreement, signed in the Achnacarry, and the US symbiotic desire for control of oil can be revisited in the late 1950s. At this time the bountiful (and cheaper) oil from the Persian Gulf basing point was transported through the Suez and crossed the Atlantic to the markets in the eastern shore of the United States—a region that ordinarily must follow the basing-point pricing at the Gulf of Mexico. At this juncture, Soviet surplus oil also found its way to the cartel-dominated world oil market, and predictably added to the drama of US "national security" and led a bumper crop of self-justifying literature that can only be described as the "Russians are coming," particularly, in the wake of Nikita Khrushchev's 12-day visit in the September of 1959 to the United States.

The flow of Soviet oil was indeed a pretext for masking the real transformation that had been underway and by then gaining further strength little by little, market by market, and region by region across the dominion of IPC. This was the rise of oil companies, known as "independents." The deliberate entrance of these oil companies, which were unrelated to the IPC—combined with leaky boundaries of carved up oil regions—had violated the cardinal rules of Achnacarry in the battleground of capitalist competition that ultimately put an end to the IPC in 1972. This was the main show as we have observed its culmination in little over a decade hence, in the oil crisis of the early 1970s. The side show was the threat of "Soviet oil," which took center stage and was choreographed for US propaganda, and which appeared more entertaining. In reality, the feeble Russian-oil explanation was a catalyst invoked by the internal undoing of IPC itself. Yet, the Russian-oil scenario is still so seditiously hot that Yergin (1991: 519–20) appears to act as a conduit to IPC—notifying us: "[T]he only way open to the companies to cope with the general oversupply and, in particular, to counter the Soviet threat...was the competitive response, price cut." This should not be lost to astute readers who recognize the context and treat the alleged "Soviet threat" on par with the "threat" by its formidable counterpart—namely, US independent oil companies—against the neocolonial and cartelized oil. This attempt at distraction though caught up with the United States in advents of the 1970s. To prevent any disturbance against cartelized control of oil, the Eisenhower administration decided to impose a quota on imported oil from the Persian Gulf, which also had some consequence for Venezuelan oil. All this was of course accomplished under the convenient rubric of "national security," which was a great invention then for letting US Justice Department off US State Department's back, and which had made the many who worked diligently in the former shudder at this audacious double-dealing.

Another instance of US appeal to cartelized control, albeit disguised as oil security, was the formation of OPEC the following year, 1960, in Baghdad. The combination of US quota and decisive cut in "posted price" of the Persian Gulf oil region was played as a one-two punch across the face of the countries (Iran, Iraq, Kuwait, and Saudi Arabia) whose oil royalties were dependent upon the price and quantity of output from the Persian Gulf. Venezuela also joined in solidarity and for the fact that the production cuts has severe effects on export of Venezuelan oil to the United States. Formation of OPEC was provoked by an arbitrary reduction of the "posted price" of most bountiful oil in the world, a critical move by IPC to counter the ensuing interregional competition.

This was a double-barrel onslaught against the owners of oil deposits in the Persian Gulf (through royalty reduction) and contrary to the interest of nearly all consumers in the United States who paid more for oil and oil products through the imposition of this quota. Hence the origin of OPEC was set off not to go up against the United States but to contest the arbitrary action and, much later, the colonial excesses of the IPC. Against this backdrop and in view of the mountain of readily available sources and US government documents, there are unambiguous records of US (and UK) opposition to creation and (formal) state recognition of OPEC. These and other involvements are a concrete example of territoriality (and absorption) of state vis-à-vis imperialism. These records also account for infirmity of US vision, US fascination with colonial control of oil. They show that the United States and United Kingdom did not acknowledge OPEC until finally forced to recognize it in mid-1960s, when the altered and exacting reality of the then international polity smacked them hard across the face with much humiliation (see Ch. 4 in this volume). And if this does not add up to an objective propensity for change, we have no idea what objectivity means.

One may look over yet another remarkable intimation of control camouflaged as concern for security of oil during the oil crisis of the early 1970s. This is a complex case that needs careful untangling of the Arab-Israeli war (October 1973) and the crisis of decartelization and globalization of oil. Although these two events appear to have been entwined, they had but little to do with each other; the brief oil embargo against the United States (and the Netherlands) was indeed a symbolic gesture against predisposed and provincial US foreign policy in the region by certain Arab oil exporting countries, among which stood one of the two formidable "pillars" of the Nixon Doctrine in the Persian Gulf—Saudi Arabia. Another "pillar" was the late Shah of Iran, who had stepped aside from this fictitious, in-house confrontation and reaped the benefits, while also supplying oil to Israel. The quarrel was not really about the US support for Israel in the conflict but about the excessive support from the United States that led to tacit encouragement of keeping and building on Palestinian (and the Arab) lands, beyond the 1967 armistice, by Israel (for a three-volume UN resolutions on Palestine and the Arab-Israeli Conflict between 1947 and 1986 see Tomeh 1975, Sherif 1988, Simpson 1988). But the embargo was not supposed to last, and it did not, as the Saudi's attitude, toward this and other issues surrounding the decartelization of oil, was directly put to the test as was demonstrated in chapter 4. Therefore, one must refrain from being sidetracked by the effect of oil embargo, so that one may focus on the real cause of the embargo, which had nothing to do with oil. Therefore, pointing to the 1973 oil embargo and speaking of a need for oil security is not unlike the story of a man who shot himself in the foot and wondered why it hurt, and then blamed his bleeding foot. This is how US foreign policy during the Nixon administration had been perceived from the outside and by the people in the region.

Even if one gives full credence to the kid-glove oil embargo of this period, one ought not steer the conversation away from the 1973–74 oil crisis, which was not complete, and which had already been set off by formidable internal forces beyond politics of any kind toward a widespread restructuring of oil across the world. Thus concern for the security of supply, triggered by internal or external contingencies, and in the period of crisis, is not out of the ordinary. Such concern for oil security is frequent, which precisely necessitates provisions like that of the US Strategic Petroleum Reserves. But this is far from alleging *security* and contemplating *control* with reference to a fully capitalistic sector that had already excelled beyond decartelization and reached its pinnacle of (competitive) globalization. Today, is it not fanciful, if not downright silly, to speak of oil security when one may secure it by simply driving oneself to an oil spot or futures market? One may ask: Is this security-mongering business not motivated by a desperate sense of nostalgia for good old neocolonial days, wherein the image of umbilical cord of the US foreign policy had to be concealed under the disguise of Cold War in broadcast media; the days of post-1953 coup d'état in Iran, for instance, when Mossadegh (an antimonopolist/anticolonialist man of democratic genre) had been ousted and dragged to the Kangaroo courts to be adjudicated by the US (and UK) surrogates in Tehran?

The 1973–74 oil crisis was a historical marker, a preamble to competitive globalization of oil and the decline and fall of American (global) hegemony. This crisis though had to resolve the question of the landed property (and ownership of subsurface deposits) in the presence of capital in valorization, wherein the oil rent in our contemporary economic system finds prominence. Oil rent thus attained the status of a category irrespective of whether OPEC had existed or not. Oil rent had to be disbursed to the owner of oil-in-place, public or private, irrespective of membership in OPEC or Rotary Club. Yet, in this turbulent transition—that is, the change-over from colonial and semicolonial oil royalties under the old concession system to postcolonial oil rents based on modern-day oil contracts—OPEC's uneasiness and corresponding challenges had not been entirely out of the ordinary. Nevertheless, there was a deliberate attempt to single out OPEC as the face of the crisis and to confront it in the name of security by creating an association for industrial countries that imported oil. The following passage is instructive:

> [O]ut of the rancorous Washington energy conference emerged the International Energy Treaty of 1974. It outlined a new energy security system that was meant to deal with disruptions, cope with crises, and avert future bruising competitions that could destroy an alliance...At the same time, it was meant to serve as a deterrent against any future use of an "oil weapon" by exporters...The treaty established the International Energy Agency (IEA) as the main mechanism for meeting these objectives. The IEA was also meant to provide a common front for the industrial countries and thus counterbalance OPEC, the Organization of Petroleum Exporting Countries...As such it [i.e. the IEA] operates as a kind of "energy conscience" for national governments. (Yergin 2011: 270–71)

By insulating OPEC's alleged offensive from the inner reality of oil, and infusing it with the ongoing oil embargo, the underlying cause of this "offensive" (other than the war) and the context of oil toward globalization were lost to the naïve onlooker. This created an opportunity for US (and UK) policy makers to revisit the old issues and purported remedies at the inception of OPEC, and to redraw a nearly identical sketch for the newly constituted International Energy Agency (IEA) in 1974. To be sure, creation of the IEA is not a novelty, as Yergin (2011: 270–73) reports, but the brainchild of the 1964 US-UK Memorandum that, among others, proposed to encircle OPEC by creating a consumer front (see Ch. 4 in this book).

Thus this is an old wine in a new bottle that is not quite informed of the objectivity of decartelization and globalization of oil, and the real reasons behind the collapse of the "postwar petroleum order" as Yergin (2011) would like us to believe. The original motivation behind the creation of IEA speaks lucidly and vociferously on its not-so-secret political and ideological objective to tame OPEC—a perceived effect and impression of globalization of oil. Ironically, however, such overt territorial concerns had little justification and even less tangible function in the presence of the decartelization and globalization of oil. Such vestiges of territoriality had no room to maneuver beyond its epoch and dwindled away with the collapse of Pax Americana by the end of the decade. And once this critical point is grasped that in the absence of the postwar system of Pax Americana virtually all paraphernalia (including NATO, the World Bank, and IMF), pretenses, pacts, practices, and policy prescriptions are passé, the significance of IEA could not be more overstated. Such redundancy, which also speaks of the absence of the trump card of IPC, is but a veritable symbol of absence of US hegemony (see Bina 1993, 1995, 2006).

To be sure, some challenges against the IPC (through OPEC) had come from unlikely quarters that had long been in cahoots with US foreign policy, and certainly in sync with Nixon's Doctrine in the Persian Gulf. Again, King Faisal of Saudi Arabia and the Shah of Iran where the two "pillars" of US policy in the Persian Gulf and they would have done whatever it would have taken to "secure" the oil for the sake of the United States. But "security" is no more than a code-word for cartelized control; it is nostalgia and thus an anachronism in the global economy and global polity that revolves around the globalization of oil since the 1970s.

As has been demonstrated, the lingering attitude about security of oil has less to do with oil and more to do with the lingering visages of the now defunct Pax Americana. These visages had made the old hegemon orphaned and allowed it to suspend between the departed and newborn global polity. To be sure, security of oil is important on its own the least of which has to do with claims that propel code-mongering and promote predisposed nostalgia about oil, and ultimately a need for US intervention. Concern for any alleged oil security cannot be without a context. The context, as we argued in this book, had lost its historical authenticity with the collapse of Pax Americana, yet it retained its pretense as a feature of US reaction to its lost hegemony. And, in a nutshell, this is the real story of oil security—a pitiful parody of our time.

## Concluding Remarks

We have argued that Lenin's era of imperialism is not the highest stage of capitalism. On the contrary, Lenin's Imperialism was the budding stage of capitalism in which capitalist social relations were still undeveloped. Conspicuous incidence of cartels and thus the absence of adequate competition was one fact of this stage. Another facet was the presence of an overwhelming majority on the planet who had little or no experience as to what capitalism was all about. Our theoretical demarcation is Marx's theory of capitalist competition, not the folksy connotation of "rivalry" in a blanketed rendition. This, we believe, was not appreciated by Hilferding or by Lenin. But our empirical demarcation, that is, our study of one of the most notorious cartels in recent memory, is more revealing. We have demonstrated that a cartel of this size, and backed by the greatest imperialist powers on earth, could collapse under its own weight, not for the lack of volition but because of the inner nature of competition. Lenin's era was an undeveloped stage of capitalism in which cartelized enterprises were the norm and capitalist competition had not yet been fully developed. And globalization of oil is merely an indirect facet of mature capitalism in which competition is fully developed.

It is remarkable that the modern economics of today is at loss as to what capitalist competition is. It sets up the fable of "pure competition" and pulls out from it a point of departure to make it more realistic. But this realism by virtue of being contingent upon a fable is a realism of a particular kind, namely, Procrustean—a heartbeat away from idealism. But this is exactly what mainstream theory does to generate "imperfect competition." This abandons the reality of competition in conjunction with *concentration* and *centralization* of capital. Here, it has been shown that decartelization and subsequent globalization of oil, since the 1970s, is a remarkable example of *real competition*—a kind that reveals the universal characteristic of capitalism. The evolution of oil is also a revolutionary example of the formation of differential oil rents (both for OPEC and non-OPEC producers) entwined with competition and competitive profits across the globe. This evolution is the consequence of the structural transformations that led to decline and fall of the colonial oil concessions in the Middle East, Africa, and Latin America, together with the entry of "independent oil companies" that eventually broke the back of the IPC (1928–72) from within and from outside. As a result, the controversy over the barrier to entry is demystified in this transformation. In theory, this chapter presents a challenge to mainstream economics orthodoxy, and good part of heterodoxy—including the radical left, on the thesis of monopoly and universality of capitalist competition. On the political plane, it demonstrates that the era of Lenin's imperialism was the era of infancy of capitalism—not its graying age. On policy, this chapter attempts to provide a hint or two about the decline and descent of the United States from the hegemonic throne.

# 6

# The Globalization of Energy

*There is no such thing as* "green energy"*—at its core, energy has neither color nor taste.*

—Anonymous

## Introduction

The purpose of this chapter is to show that, since the early 1970s, the market prices of all energy sources have been regulated by the global market value of crude oil, which in turn depends upon the magnitude of individual value (or production price) of production units located within the least productive oil fields of the continental United States. Thus, historically and methodologically, the globalization of the petroleum industry, together with the preponderant influence of US oil capital over other energy sources—especially its control over US coal—should be considered as (1) the primary basis of an all-embracing energy industry consisting of all sources and (2) the driving force behind the globalization of the energy industry across the world. In this fashion, the individual capitals associated with the production units belonging to traditionally autonomous industries (such as coal, oil, and natural gas industries) had to compete directly with each other regardless of the immediate use value produced, according to the law of intraindustry rather than interindustry competition. Hence, contrary to the conventional wisdom, the control of oil capital over other energy sources, which resulted in further integration of energy, has led to further competition in the sphere of production within the entire energy industry worldwide.

From the mid-1800s on, coal became the primary source of energy in the United Sates. In Europe coal had gained importance nearly a century earlier following the Industrial Revolution, and became the lifeblood of the early development of capitalism. It was not until the end of the Second World War that the importance of coal diminished as it was gradually replaced by crude oil and its joint product, natural gas, as cheap sources were suddenly discovered in the vast oil fields of the Middle East. Nevertheless, as the consumption of coal by both the industrialized countries of the West and the emerging economies of the "Third World" continued, its more traditional applications were gradually diminished, leaving room for the newly built economic structure to be largely fuelled by oil. In the meantime, given the demand for more energy that fueled the transformation of capitalism in the postwar period, in addition to consumption of fossil fuels, energy production from other sources, such as nuclear, hydroelectric, solar, and wind-power, has also been increased. However, all these sources combined have not yet produced sufficiently to influence the magnitude of energy prices (or values) in the energy market. Energy from fossil fuels is by far the most dominant form of energy in use today.

However, there are three essential sources of energy in the fossil fuel category, namely, coal, oil, and natural gas. The importance of the three sources—that is, coal, oil and natural gas, whose combined share is nearly 90 percent of the energy demand worldwide—cannot be underestimated (Bina 2012a: 154, Table 1). Since 1947, the supply of these three types of fuels combined has been more than 95 percent of total energy until the late 1970s (EIA 1982b: 6). However, there has been a steady but slow decline in the combined share of these fuels to 83 percent in 2009, while projection for 2030 is 78 percent in the United States (Source: EIA, *Annual Energy Review* 2012). This means that, in the face of various changes in the structure of the energy market, the share of these fuels combined has remained stable for some 30 years *(Coal Data Book*, GPO 1980: 15.) There has been a substantial change, however, within this class of fossil fuels that led to a significant decline in the share of US coal production, from 46 percent to nearly 20 percent during the 1947–77 period; whereas the share of oil production increased from 37 percent to 49 percent during the same period. The results for the combined share of crude oil and natural gas are even more significant. The trends were stabilized by the first decade of the twenty-first century for the United States, showing correspondingly 21 percent, 37 percent, and 24 percent for coal, oil, and natural gas for 2009; the projections for 2030 are not significantly different for the shares (Source: EIA *Annual Energy Review* December 16, 2010).

A similar trend exists for the entire world energy market. During the period 1950–75, the share of world coal production declined rapidly, from nearly 60 percent to about 34 percent, while the oil production share increased from 29 percent to nearly 46 percent of the total fossil fuel production (EIA 1982b: 5). From the standpoint of world energy demand, we observe, more or less, the same pattern of shifting from coal to oil. Given the economic, political, and technological developments during the period 1928–78, the share of coal in the world energy demand fell sharply from 74 percent to just 27 percent (Stock 1980: 208). During the same period, the combined share of oil and natural gas rose from 19 percent to 65 percent (Bina 2012a: 154, Table 1). This trend is stabilized similarly for the same period and with the same pattern as that of the United States (BP *Statistical Review* 2012).

## "Commodity Energy" and Formation of Energy Industry

Based on neoclassical price theory, having a unified pricing mechanism for all energy sources depends essentially on the existence of: (1) an adequate conceptual framework for a multiproduct industry that provides a uniform competitive market for all energy sources, regardless of their physical form; (2) a derived demand for energy, in common physical units, that results from aggregation of the intermediate demands for all sources, without having aggregation problem; and (3) a high degree of substitutability among these sources, with negligible conversion and transmission cost. Among many conceptual difficulties pertaining to these points, there is a qualified consensus among orthodox economists that "there is relatively little substitutability, with oil in the bulk of its current uses, especially in terms of short term conversion" (US Congress, *Petroleum Industry Involvement in Alternative Sources of Energy*, 1977). The above conditions will become further complicated where the entire matter will be subject to the global market and conventional arguments surrounding the perceived cartel and monopoly conditions. Therefore, there is hardly any reliable methodological (and theoretical) framework, other than a patchwork of axioms, to respond to this very empirical question without entanglement in data crunching, simulation, and speculative reasoning.

Alternatively, with respect to Marxian value theory, there is no ready-made, satisfactory theory for the unification of all energy sources within a unique value-theoretic framework. In this chapter, we attempt to develop a theoretical framework based on Marx's value theory (and his theory of competition) for the entire energy industry. This attempt shall be made in light of the contemporary debates on value theory and theory of competition within Marxian tradition.

It has been argued that comparison of various energy sources will become possible if these sources can be converted into a common material form capable of being measured in physical units such as calories (Massarrat 1980). It has also been suggested that, since among the carbon-bearing sources of energy coal and oil are the most significant, the market price of one fuel has to regulate the price of other fuels, and also prices of all remaining sources of energy (Massarrat 1980).

One way to approach the above problem, it is alleged, is to compare equal amounts of coal and oil in appropriate physical units, such as British Thermal Unites (BTU) or calories, and to determine which one on average requires a greater mass of capital for its production. Thus, the larger the mass of capital, the higher the corresponding value. This framework explicitly suggests that "Because of its particularly unfavorable use-value form, a much greater mass of capital is necessary to produce a given mass of energy in the form of coal than is required to produce the same mass of calories in the form of crude oil" (Massarrat 1980:34).

The following passage explains how Massarrat had arrived at his theory of price formation for energy industry based on production price of coal. He simply concludes:

> The productivity of labor in the production of crude oil is, therefore, several times higher than in the production of coal, as the material basis of the former is more favorable than that of the latter... Thus the individual price of production of coal regulates the market price of all other carbon bearing use-value forms of energy sources. (Massarrat 1980: 34)

Clearly, the above approach is a contrasting alternative to conventional neoclassical pricing and policy proposals that would inevitably come about from it. This is certainly a timely contribution in view of contemporary renewal of interest in the classical and Marxian political economy. Nevertheless, it is equally clear that this method of approaching the value theory would inevitably reopen a Pandora's box of some well-known controversies surrounding the theory of capital (known as Cambridge capital controversy) within both Marxian and neo-Ricardian economics (see Harcourt 1972; Marx's *Capital* vol. 3: 1981, Part II; Fine 1982a; Shaikh 1982; Steedman and Sweezy 1981; and Mandel and Freeman 1984). With this in mind, we review the theory of energy pricing presented by Massarrat 1980, before offering an alternative theory with quite different conclusions about value and pricing of energy sources.

First, given the ownership and penetration of oil capital into the production process of other sources of energy, especially production of coal, one has to allow for a direct, face-to-face contention, via the process of intraindustry competition, between the various individual producing units associated with all the major energy sources. For instance, individual capitals within the coal industry should be able to compete *directly* with those of the oil and the natural gas industries. In other words, the traditional concept of industry based on particularity of use value has now been transformed into a concept that, in conformity with the centralization of capital, unifies the various production processes under one unified industry and one rule. This leads to amplified competition among individual capitals, associated with different forms of energy, which otherwise were separated by industrial boundaries and which were only allowed to compete beyond the realm of a single energy form in interindustry competition. Centralization of capital in the energy industry under control of oil capital treats all production units in coal, oil, and natural gas in direct competition within the same industry. As these individual energy-producing capitals compete against each other directly, they also form a unique market value in conjunction with different profit rates that result from the productivity (and profitability) of each individual production process, regardless of their immediate use-value forms (see Marx 1981, Ch. 10; Fine 1982b, Ch. 8; Shaikh 1982; Bina 1985, Chs 6 and 9; Bina 1989b, 2006).

Second, the question of "labor intensity" (or "capital intensity") must be treated on the basis of a clear distinction between the *material* side and the *value* side of production and exchange process based on the distinction of use value from exchange value, which has critical implications for a classical political economy and its celebrated Marxist critique (Marx, 1970, Ch. 1; Marx, 1981, Part III; and Fine, 1982b, Ch. 8). The definition of capital/labor-intensiveness used in modern economics literature is problematic and often misleading in light of the above consideration if, following Marx's recognition, one does not know whether the above concept refers to material side or value side of the commodities. In fact, modern neoclassical economics removed this distinction in its framework by assumption of the identity of value on the exchange side and value on the material side, thus opening a well over a century-old controversy based on the lack of commensurability on the part of use value based on individual utility.

From the standpoint of classical and Marxian political economy, however, labor-or-capital intensity has to be measured through the value side alone, as the apparent forms of the labor processes and their material manifestation would not provide any indication for measurement of value—a priori. In Marxian economics, if the relative value of constant capital to that of variable capital in a commodity vis-à-vis the relative value of constant capital to that of variable capital associated with the *social average* is taken to measure the capital-or-labor intensity (i.e., a comparative assessment of the organic composition of capital associated with each process), then a seemingly labor-intensive process, from the standpoint of its material side, can be easily viewed as a capital-intensive process according to its value side.

Third, if the above conceptual framework is applied to the energy industry as a unified whole, then the individual producers belonging to coal, oil, and natural gas industries tend to compete face-to-face alongside one another, regardless of their industry of origin, within the realm of the newly emerged energy industry. The ultimate use value here is *commodity energy*, reflecting the material side of production, whose value is *regulated* by the individual value of the capital that is extended to the least productive value form, which, a priori, may belong to any of the above (use values) sources. In other words, given the above framework, since there is no longer a necessity to presuppose the formation of separate market values for coal, oil, natural gas, or other forms of energy, the existing individual capitals associated with the production of various sources of energy must be treated on the same footing. The competition of these individual capitals can well be categorized as intraindustry competition, which simply precludes these sources (i.e., use-value forms) from acquiring a separate market value. To put it differently, one cannot possibly argue for the development of an all-embracing sui generis industry that has a unique market value for all energy sources and, at the same time, treat these sources as if they belong to different industries, each with a market value of its own and then imagine the unimaginable that it would be possible to select the highest market value that regulates the value of the entire energy industry. This theoretical conundrum, that is, mutual exclusivity of having an energy industry and also wanting to treat all forms of energy as if competing along an interindustry trajectory is not only theoretically bizarre but also misleading from the standpoint of energy policy. This shows the richness and the range of application of Marx's value theory, particularly his theory on competition, to mature capitalism of today. Moreover, we learn in this context that the distinction of intra- and interindustry competition is not only meaningful as to the formation of value and production prices of energy but also handy for bringing to light the theoretical basis as to why this industry is also a global industry.[1]

To be able to treat all the major sources of energy within a unified and all-embracing industry systematically, one needs to consider having a spectrum of competing energy-producing capitals consisting of individual producers from all major energy sources, arranged according to their productivity rather than their immediate use value (i.e., coal, oil, natural gas, uranium, hydro, geothermal, solar, tidal, and wind, among others). Without a priori assumption, in actuality one may find any number of coal-producing units (i.e., individual capitals) that are either more or less productive than those of their oil-producing counterparts are—within the above spectrum. Similarly, one may also find particular coal-producing units that are either more or less productive than their natural gas counterparts are. The key point here is to stress the fact that it is indeed possible (and we will demonstrate that it is actually true) to have individual coal-producing units (capitals) that are more productive than their oil-producing counterparts, even though there are many other oil-producing operations that in fact might very well prove to be among the most productive oil producing units within the energy industry. Thus, one cannot a priori consider a higher or lower value magnitude for coal, oil, or other sources of energy by merely comparing the values of *average* producers from each category.

Fourth, since the structure of value production in the energy industry is influenced, more or less, by the structure of property relations that would manifest itself through the formation of rent, the existing differential productivities translate into differential profitability, on a permanent basis, within the entire industry. Hence, some producers tend to gain surplus profits in the form of differential rent, a sum of money to be appropriated by the owners of mineral deposits, which is utilized in the production of energy. Thus, it is appropriate to refer to such rents as *differential energy rents*, regardless of the specific forms of such deposits.[2] We have chosen significant sources within the fossil fuels category, a combined volume of which by far outweighs both the production and consumption of the remaining sources of energy combined, both in the United State and the World. Our selection of two out of the three above-mentioned major energy sources here (i.e., coal, oil and natural gas) is also not arbitrary. In fact, we dropped natural gas because, as a joint product, its value is essentially dependent on the market value of crude oil. As a consequence, it would be sufficient to include only coal and oil in the analysis of price formation for the entire energy industry.

## TECHNICAL CHANGE AND COMPOSITION OF CAPITAL

Organic composition of capital (OCC) is simply the reflection of technical composition of capital (TCC) in terms of existing (hitherto established) values whose magnitude is due to be adjusted to a new level with different value composition of capital (VCC), following the restructuring of capital by way of periodic crises (Marx 1981, Part III; Fine and Harris 1979, Ch. 5; Fine 1982b, Ch. 8; Fine 1986; Shaikh 1978, 1984, 2010; Saad-Filho 1993; Bina and Davis 2008).

The new VCC is the outcome of the change in the TCC and its manifestation in value terms within the unity of production and exchange. Thus, the understanding of OCC is contingent upon the understanding of the changes on the material side, albeit in value terms, before the formation of a new value with a different composition of capital. The above process necessitates the recurrence of a periodic break in the circuit of capital that signifies frequent discontinuity in the production of value whose magnitude is subject to change through periodic crises. Economic crises, therefore, are the outward manifestations or symptoms of the restructuring of capital within the accumulation process affected through periodic innovation and change in the material conditions of production in capitalism—the veritable replacement of variable by constant capital. Parenthetically, economic crises do not automatically lead to breakdown of capitalism, as many on the popular and radical left like to believe. Economic crises are a mechanism for restructuring and renewal of the circuit of social capital. Crises could hardly break the system, people do—class struggle does.

Returning to the main subject, to be able to account for the value side, following Marx's own formulation, we have to recognize the two different impacts that can result from technical change, that is, change in the material conditions of production—TCC: (1) a direct effect that leads to a change in OCC in conjunction and within the same value magnitude; and (2) an indirect influence that, overall, changes the magnitude of value (Fine 1982b, Ch. 9; Bina and Davis 2008). Thus, according to Marx, the economic crises in capitalism cannot be adequately theorized without the three-way distinction among TCC, OCC, and VCC. Revisiting of the following passage from *Capital* is instructive:

> The composition of capital is to be understood in a twofold sense. As *value*, it is determined by the proportion in which it is divided into constant capital, or the value of the means of production, and variable capital, or the value of labor-power, the sum total of wages. As *material*, as it functions in the process of production, all capital is divided into means of production and living labor-power...I call the former the value-composition, the latter the technical composition of capital...[In addition,]...I call the *value* composition of capital, in so far as it is determined by its technical composition and mirrors the changes in the latter, the *organic* composition of capital. (Marx 1977 [1867]: 762, emphases added)

The above distinction is methodologically in conformity with Marx's method of starting with the fundamental categories rather than individual agents or micro-behavior (Fine 1982b, Ch. 9). Because of the rise of TCC, that is, an increase in the volume and mass of constant capital per unit of variable capital, OCC will increase. This increase is the characteristic of *technical change* and its immediate manifestation within the sphere of production, prior to the formation of newly emerging value with a smaller magnitude and different composition of capital. Such a technical change, therefore, leads to higher labor productivity and lower value magnitude per unit of commodity, a phenomenon that is also known as the *cheapening* of commodities.

Finally, while on the subject, it is important to point out that the reproduction of capitalism, as a social system on its own, does not essentially depend on "cheap labor." The latter is a demonstration of rapid suspension of precapitalist conditions and the conquest of encroaching capital. Cheap labor is the result of immediate application of capital—a precondition created by social capital for the purpose of autonomous and sovereign propagation. Hence cheap labor is necessary but not sufficient for the survival of capitalism. Cheap labor, however, views capitalism from the prism of *individual* capital. This also goes for other "cheap things"—like "cheap oil." The essential mission of capital is the *cheapening* of labor power—the mother of all commodities. And this is done through incessant technological change.

## Centralization of Capital and Regulation of Energy under Oil

Aside from the physical and material conversion of energy sources into a common use-value form, we have maintained that theoretically "commodity energy" should have a value whose magnitude is dependent on the value (or production price) of either coal or crude oil. At the same time, we have either demonstrated or implied that (1) in a sui generis energy industry intraindustry competition requires that all production units (i.e., individual capitals), regardless of the physical form of their output, have to compete with each other directly; (2) the mass of capital associated with a particular energy source would not necessarily tell us anything about value dimension; (3) the concept of labor intensity must be treated on the basis of value composition rather than material composition of capital; (4) based on the existing differential productivities in the production of coal and oil, as major sources of energy, one can find production units (firms or individual capitals) with larger or smaller individual value that produce one or both types of fuels (without any particular arrangement based on their industry of origin) across the energy industry; there is no a priori reason to assume that coal production per se is more expensive (or less productive) than that of oil; and (5) the control of oil capital over the production of coal (and other sources of energy) is a precondition for further competition through *centralization* of social capital.

The extent of centralization of capital through the control of the US oil industry over US coal is significant. In addition, we have to realize that more than 40 percent of privately held US coal reserves belong to US oil and gas companies and their subsidiaries (Bina 2012a: 161, Table 2). This means that the individual oil-producing as well as coal-producing companies have all been collectively influenced by oil capital as if they were members of the same industry. For more information pertaining to the control of oil capital over all other sources of energy, including the known US uranium reserves see US Congress 1977 and Chapman 1983.

It is apparent that, if we follow Massarrat (1980), a simple comparison of energy sources from the standpoint of their embodied mass of capital remains more in tune with the labor-embodied Ricardian value theory than the socially necessary abstract labor in Marx. As we know, the former has grappled with a number of insurmountable complications even within its own framework. Since the regulating value (or production price) in the energy industry may depend on the individual market value of any particular energy form, together with the process of centralization, the historical significance, shared relationship, and the gradual displacement of one energy form by one another must be empirically investigated. That is why we have to attach great importance to the role of oil capital in the regulation of energy prices through the control of coal (and other energy sources) in creation of a sui generis energy industry, beyond any one nation, across the global divide today.

Empirically, it is interesting to note that the price of US coal, in terms of BTUs, has always been rather substantially below the US oil price during the period 1951–81 (Bina 2012a: 163, Table 3). At the same time, coal prices increased, during the oil crises of the 1970s, so suddenly and significantly as to suggest they were pegged to oil prices ( Bina 2012a: 163, Table 3). However, the comparison of the US coal and oil capital expenditures during the period 1971–75 does not only support our main thesis, but also distinctly illustrates that the long-term cost of coal was only a fraction of the cost of oil in the United States (Bina 2012a: 164, Table 4). Thus, contrary to Massarrat (1980), we conclude, on both theoretical and empirical grounds, that the market value of oil should regulate the value and prices of all sources of energy, fossil and nonfossil, renewable or nonrenewable—including value and price of coal.

As for the market value (or production price) of oil, as was demonstrated in this book, the individual value of the least productive US oil fields has become the regulating market value of the entire oil industry worldwide (Bina 1985, Chs. 7 and 9, 2006). This means that the regulating capital linked to the US domestic oil not only regulates the global production price of oil but also the global production price of energy in all its physical form across the globe. This has been the result of globalization of capital in the oil industry, which in turn has led to the unification of global oil production under a worldwide pricing mechanism (Bina 1988, 2006, 2012a). Thus, given (1) the globalization of oil, and (2) the control of oil capital over coal and other sources of energy (this includes ownership of the vast coal, natural gas, and uranium reserves by oil companies), we conclude that the market value (production price) of oil regulates the prices of all energy sources anywhere on the planet. Moreover, the globalization of energy is a direct result of the globalization of oil, which is an event of specific imperative in our time. Therefore, there is hardly any policy question in any field of fuels in the world that could be exempt from the fact that the US oil alone has a profound effect on the momentum and direction of future energy.

The above conclusion is in stark contrast with that of Massarrat 1980: 32, which asserts "It is clear that the energy sector must be seen not on the national level but internationally, since the market price for the product of this sphere, like the product of all spheres of production immediately dependent on nature, can in general only be derived in an international context." First, it is both theoretically without foundation and empirically unproven to argue that the commodities whose production *is* immediately dependent on nature should be necessarily priced internationally. It is rather odd to consider "nature" as a cause of internationalization. Besides, a theory of rent that is fundamentally based on nature is an ahistorical construct to be counted as a part of social relations in capitalism. Such a theory also departs from Marx's notion of rent (Bina 1985, Ch. 5, 1989b, 1992, 2006). Second, the internationalization process is historically a specific stage in the self-expansion of capital that signifies its centralization on the world scale (Palloix 1975, Bina 1985, Ch. 9). We shall attend to this question more fully in the following section.

Aside from the nationalized coal industries in Europe, the market value of oil regulates the prices of US coal, natural gas, and all other sources of energy worldwide. Thus, the value of "commodity energy" has to be identical with and regulated by the market value of global oil. This latter value is none other than the value of US domestic oil that, in turn, is regulated by the value of oil from the least productive US oil fields. The capital linked to the latter oil fields is the regulating capital of the entire oil industry and, by implication, also constitutes the regulating capital for the entire global energy sector. Consequently, the market value of oil tends to regulate the prices of all energy sources globally. Given the above symbiotic relations among all energy forms, the oil crises are no longer limited to the realm of oil alone, as they evolve both across the various forms of energy and transverse the entire geography of production worldwide. That is how restructuring of capital in the oil industry can cause the restructuring of the entire global energy industry.

## Oil Rent and Other Outstanding Questions

In this section we attend to several issues of critical import related to Marx's rent theory and Massarrat's interpretation concerning the fossil fuels and energy. We begin with a brief reexamination of Massarrat (1980: 26–68) and in doing so will provide a brief critique of the works by scholars who were influenced by Massarrat's vision of rent and value theory.[3] First, the point of departure in Massarrat is "commodity energy," an abstract category, which itself is a derivative of the concrete forms of energy sources. Massarrat's point of arrival is also the same "commodity energy," hence, the simultaneity of the premise and the conclusion (1980: 32–35). Second, as has been pointed out earlier, Massarrat begins with the unfavorable use-value form of coal vis-à-vis that of oil, comparing a *hypothetical* ton of coal with its crude oil equivalent in terms of their calorie content. He then contends that the "productivity of labor [is much higher] in the production of crude oil" than that of coal in this *hypothetical average*, before concluding that "the individual price of production of coal regulates the market price of all other [sources of energy]" (1980: 34). Why is this procedure faulty? Because, such reasoning stands on shaky grounds by the comparison of hypothetical averages that is susceptible to circularity. Also, as has been pointed out above, the premise of higher price per calorie for US coal is empirically untrue for the period under study (Bina 1989a: 167, Table 3). Third, Massarrat tends to shift the problem from the cause of the oil crisis to its effects by focusing on coal. There is neither a concrete empirical analysis of US oil nor a concrete theorization of the evolution of landed property in oil nor even any analysis of US coal, which is supposed to be the focus of his thesis (Massarrat 1980).

Fourth, as was pointed out earlier, the most serious error in Massarrat (1980) is the conflation of the levels of analysis in the construction of "commodity energy," namely, the utter confusion of the intra- and interindustry competition. The requirement for an adequate analysis is thus to start with the real site of the crisis—oil. This entails a thorough examination of the globalization of oil, valorization of the oil deposits via the landed property associated with them in the United States, the Middle East, Africa, and Latin America (i.e., OPEC and non-OPEC oil), formation of differential oil rents, and lastly the rationalization of US oil industry all in one breath within the 1973–74 oil crisis itself (see Bina 1985, 1989b). This way a concrete examination, which is informed of its theoretical limits and vice versa, would be free from speculative and axiomatic reasoning.

Finally, Massarrat appears to have relied on a static interpretation of Marx's AR as monopoly. In this case, why should the magnitude of AR as a monopoly rent be prearranged within the limits of market price and price of production? Why would this "monopoly rent" not be like any other ad hoc monopoly, being so fixed arbitrarily? As we have shown in this volume, the answer to these questions is quite simple: Marx's AR is not a monopoly rent but rather a rent that is reflective of the effect of obstacles of landed property against the flow of capital onto agriculture. That is why Marx specifically speaks of OCC as he precisely identifies the limit of AR's magnitude as the difference between value and price of production—a so-called maximum toll in agriculture. As we have shown in this book, for Marx, monopoly of (valorized) landed property is synthetic, not "natural."

Given the high pace of capital accumulation in the oil sector, speaking of AR is absolutely irrelevant to the production of oil. This is a well-documented fact. Yet, Massarrat keeps referring to natural monopoly and monopoly rents in rather orthodox terms. He writes:

> Landed property excludes this natural basis [raw materials] belonging to it from production until it receives a fee. This will not happen until the social need for this raw material exceeds the supply in the long term and the market value, and with it the market price therefore rises above general price of production . . . employed in the sphere. The difference between the market price of the commodity in question and the general price of production is then (as a particular form of surplus profit, of the *natural monopoly* profit which enters into the price of commodity) transformed into *absolute ground rent* by being appropriated by landed property. (1980: 32, emphasis added)

The influence of Massarrat's interpretation of Marx on the subject of oil and rent is unmistakable. The theme of this influence can be seen in three ways: (1) by rendering of Marx's absolute rent (AR) as monopoly; (2) by a priori (value) comparison of oil and coal based on physical characteristic (i.e., use value); (3) by alleged application of Marx's AR to oil.

For instance, Nwoke argues that "The 'monopoly ownership' of Third World landlord governments would be stronger, and the magnitude of absolute rent exacted by them would be greater, as the demand for minerals expands and the supply of rich mineral deposits becomes more limited" (1987: 30). The orthodoxy becomes a bit more transparent when Nwoke "[concludes] that OPEC has only succeeded temporarily as a cartel in capturing absolute rents for its oil-exporting member governments" (Ibid. 103).

Parallel with Massarrat, Nore also begins with the axiomatic yardstick of "marginal energy producer on a world scale, measured in energy units," and concludes that such a measure must be "the basis for *the final price of oil to the consumers*." This price, he says, must be above the production price of oil but equal to the price of production of the "marginal energy producer" (1980:70–71, emphasis added).[4] This is how Nore speaks of the total amount of surplus profits in the oil industry. Now the remaining question is the distribution among the importing states, oil companies, and the producer country. What is the mechanism for such a distribution? Save for the importing states, Nore "focuses attention on a political element in the determination of absolute rent," which is now the subject of the "struggle between the owners of a non-reproducible property and the producers of commodities." Nore then concludes that "the surplus profit captured by the oil company [is] monopoly profit, [the source of a]...high level of concentration...due to its character as 'natural monopoly' and its strategic importance" (1980: 71).

Finally, Nore appears to have taken his argument, over AR, one step further and wonderfully turned the table against the "shortcoming of Marx's treatment." But, alas, a cursory inspection of all this clearly reveals that Nore's revision of Marx is not without a hefty price of self-indictment; it simply comes at the staggering cost of accepting the fallacious construction of "commodity energy," an arbitrary depiction of AR, and an orthodox vision of competition and "natural monopoly" (1980: 71–72). Also Nore's casual pronouncement of oil's "strategic importance" in the same sentence is but icing on the cake of bourgeois monopoly (1980: 71). No wonder that the scholars in this tradition had no choice other than matching their perceived monopoly with the notion of imperialism, albeit as a crutch, to plug the hole in theory by filling the insurmountable gap between alleged economics and supposed politics. Recently, Harvey (2003, Ch. 1; 2010, Ch. 3) also appears to be tagging along similar paths. The political elements, and indeed the struggle of OPEC rentier states, however, are too important to be described in arbitrary terms. The oil crisis of 1973–74 has revealed that the struggles of the rentier states over the distribution of oil surplus profits were entwined with the worldwide restructuring of the oil industry and *competitive* formation of differential oil rents across the globe. Hence these struggles themselves are neither arbitrary nor without a limit.

## CONCLUDING REMARKS

The emergence of a sui generis global energy industry depends on two de facto principles: (1) centralization of capital in the production of major US energy sources under the control of oil capital, and (2) transnationalization of capital in the oil industry and the consequential regulation of oil prices by the market value of US oil. Having dealt with both aspects of the use value and exchange value of "commodity energy," we have maintained that since the market value of US crude oil tends to regulate the prices of all energy sources worldwide, it must also regulate the price of their common material form. Thus, the price of energy depends on the magnitude of the market value of US oil.

Contrary to Massarrat (1980), we have demonstrated that, according to the law of intraindustry competition, all the individual producing units (capitals), regardless of their energy form, have to compete against each other directly, rather than competing separately based on their own physical characteristics, which theoretically demand forming a common market value for all, before entering into interindustry competition. We have shown that departure from the suggested formulation in this chapter violates the theoretical priority of having a unique market value for a sui generis industry along with different individual profit rates through intraindustry competition. At the same time, Massarrat's misinterpretation of Marx's absolute rent as monopoly and his rendition of "natural monopoly" in oil generated no less of a pseudo-problem in the context of theory and its application to politics and policy concerning oil and energy, particularly when it comes to the Middle East. Here "oil monopoly" turns into a parody that pits "imperialists" and "anti-imperialists" against each other in a context that is emptied of text.

Given the existence of specific mineral property rights, which are intertwined with the production of fossil fuels, the existing differential productivity, and the resultant differential profitability, would turn into differential rents among all energy-producing units through intraindustry competition. Since these rents are the effects of the ownership of minerals regardless of their physical forms, we identify them as *differential energy rents*. And this should provide significant theoretical insight for globalization of the energy sector as a whole. Finally, since it has been shown that the market value of oil tends to regulate the price of all energy sources, it seems consistent to consider oil crises tantamount to energy crises.

# 7

# War, Oil, and Conundrum of Hegemony

*We will get on our way to a new record of expansion... that will carry us into the next American Century.*

—George H. W. Bush
State of the Union Message to the Nation (1991)

*A foreign policy that is both immoral and unsuccessful is not simply stupid, it is increasingly dangerous to those who practice or favor it. That is the predicament that the United States now confronts.*

—Gabriel Kolko
*Another Century of War* (2002: 138)

## Introduction

The invasion of Iraq—and subsequent occupation of this little country in March 2003—is a déjà vu in the context of the repeated attempts by the United Sates at turning back the clock of the new epoch to save her hegemonic past. This unreasonable behavior has come to be reasonably predictable, particularly in the aftermath of the breakup of the Soviet Union (1917–90) and the lack of inhibition to and penchant for military adventure in the confused and confusing interlude between the collapse of the Pax Americana (1945–79) and the posthegemonic world in the making. Thanks to the loud and boisterous pageantry of the Reagan era, that masked the scale and immensity of Pax Americana's breakdown, the celebrating of the triumph of America and its acclaimed mastery of the world continued. The post-Reagan 1990s did not produce any appreciable deviation from this euphoric norm until the boomerang of 9/11 hit the homeland, not by vigorous enemies who had come from outer space, but the descendants of the foot soldiers trained and weaponized by America in the Afghan war against the Soviet "infidels"—a barefooted bunch. Even so, this was not accepted as a sign of diminishing US global power in the post-Pax Americana era. Indeed, we argue the fall of the Soviet bloc should also be considered as a subset of the fall of Pax Americana, if one cares to look at the grand trajectory of the entire forces that were brought to bear by a new epoch better known as globalization (see Bina 1997, 2011; Bina and Davis 2008). This chapter shows that the war-for-oil scenario is a misleading myth that not only negates the decartelization of oil toward competitive globalization but also conceals the fall of Pax Americana and the loss of the American hegemony. The loss of the latter and America's obsession to get it back is what this havoc is all about. Moreover, the real scenario is the decline and fall of American global hegemony, an 800 pound gorilla that has all the while been sitting in the room without an inkling of acknowledgement from the sizable majority of self-styled experts, from the far right to the far left within the political and ideological spectrum across the world. We contend that the recognition of this very key fact is indispensable for nearly all matters that affect the question of peace and war in today's global polity.

## The Mirror of Iraq and Disjointed Time

While the primordial dust of the Big Bang, so to speak, of globalization (distinct from what is known as neoliberalism), especially globalization of polity has not yet quite settled and the rules of engagement, interaction and coexistence are still in flux, the biggest bully on the block seems to be the *decider*. In this peculiar half-way scramble of politics the footings have already crumbled although the face and the façade are lingering and replicating the pathological pageantry of better times and the rumbling of reckless military might. The consequence of course is not unpredictable. The blowback arrives not in time, but in an awesome second. And that is an impeccable indication for sizing up the dilemma of a power that has lost its hegemony and now is on the descent, perhaps in free fall. The difference between an ascending power and a descending one is that the former wins its battles on its own terrain in its own *epoch*; the latter forces its purposeless battles on an epoch that negates its very existence. The first combats with legitimacy, the second confronts and crawls into a mortifying void. Hence, the hastening of the decay by the latter's own flagellating hand through exceptionalism, unilateralism, preemptive adventurism, and other assorted deceitful and reactionary means. This is the total difference when the specter of change is hovering in every nook and cranny of the world and, in a vivid Shakespearian depiction, when "the time is out of joint". That is why a descending power is more dangerous than its ascending counterpart where it comes to world peace. This is the predicament of the United States at the present disjointed time.

This is precisely why the United States, in the building of the military coalition, which had been troublesome back in 1990 was, in this second try, unworthy of its name; particularly in view of the fact that neither the region's friendliest (US) client-states nor the spirited "partners" of the exclusive imperialist club of the now defunct *Pax Americana* had much desire to join this hazy invasion. Britain though had been an aberration in this and the previous Anglo-American adventure (1990–91), which was not unrelated to the chain of events that led to the fall of the Berlin Wall and the collapse of the Soviet bloc. The mirror image of this venturesome undertaking had also been revealed in the deepening of the differences in the United Nations' Security Council, and the widening cleavage within the ranks of NATO itself. This, however, was expected, given the protracted global transition that had long been in progress in the period between the seemingly quiet implosion of *Pax Americana* in the late 1970s, and not so quiet breakup of the Soviet bloc in the late 1980s. As it turned out, the objective conditions of the emerging international polity and subjective tendencies of American unilateralism did not find mutual congenial ground on the epochal plane of globalization.[1] Loss of American hegemony prompted undisguised bellicosity, culminating in outright aggression by the Bush-Cheney administration. The war against a weak, symbolic enemy seemed inevitable.[2]

As has been underpinned earlier in this volume, the war-for-oil scenario is a misleading myth that contradicts the whole host of transformative and epoch-making changes in the evolution of today's oil industry. Any such neglect, in appreciation of oil's importance, is also tantamount to the neglect of matters that intersect with the issue of American hegemony, which in turn isolates us from sufficient appreciation of the epoch in which we live. The slogan of "No Blood for Oil" is a hoax—advertent or inadvertent. First, this slogan ignores the theoretically informed and historically endowed periodization of oil into a threefold stage of development toward globalization. Second, it overlooks the distinction between the cartelized regimes of "administered pricing" and competitive pricing according to the objective conditions and dynamics of global oil markets.[3] Third, it neglects the systemic interdependence of oil production, irrespective of the cost of production or location of oil reserves across the world.[4] Fourth, by focusing on OPEC alone, the war-for-oil scenario discounts the pivotal role of the global oil industry's *regulating capital*, which operates within the least productive US domestic oil fields and sets the production price of oil in the world. Fifth, based on a preconceived notion of monopoly, this scenario is not informed of the fact that the production price oil in the US acts as the center of gravity of oscillating market prices and that the forces of demand and supply of oil are part and parcel of such oscillations. Sixth, it is oblivious that OPEC prices are constrained by the worldwide competitive spot (oil) prices, which in turn would render OPEC (and non-OPEC) oil rents subject to worldwide competition.[5] Finally, the war-for-oil scenario does not recognize that concepts such as "access," "dependency," "control," or "energy independence" have no place in the context of today's post-cartelized global oil industry.[6]

By rejecting the epiphenomenon of the war-for-oil scenario as the cause, this chapter focuses on deeper forces of an epochal nature that brought the postwar interstate system of Pax Americana down and thus caused the fall of the American hegemony. And it is the loss of hegemony and the reaction to it by the United States that essentially underpins the cause of nearly all the post-Pax Americana conflicts, including the consecutive invasions of Iraq by the United States.[7] Moreover, we contend that the issue of war and peace, and thus stability, is inseparable from the question of US hegemonic loss, in view of the epochal changes that have redefined the nature of power in the global polity of today. And the vulgar and reductionist claim of the "oil grab"[8] can neither be squared with the globalization of oil nor can it be settled by colonial domination in the postcolonial era. Besides, domination (and direct control) is not the same as hegemony. The "blood," that is so attributed to "oil," is a mockery—a hoax; real blood is dripping from disjointed time intersected with the warped space of hegemony and its obliteration, and a declining power that is adamant to turn back the hands of time. Oil appears as a measly catalyst at best.

Well over two decades ago, in the aftermath of the 1991 Persian Gulf War, we expressed concern over the consequence of US military intervention and the latest posture associated with the presence of seemingly "victorious" US forces in the Persian Gulf. At the time, we detected a curious contradiction in the reversal of US policy toward Saddam Hussein—a sudden reversal from being friend in the long and bloody war against Iran to being a foe that had to go. Shortly thereafter, we wrote, "After all, this colossal military power was unleashed against a junior partner that was an active participant in U.S. (Persian Gulf) policy for a good number of years."

We were convinced then as we are now that neither the rhetoric of democracy nor the purported "access" to "cheap oil" was the cause of US intervention in Iraq. Rather these unilateral and semi-unilateral interventions, particularly since the fall of the Soviet Union, are the latent reactions to the implosion of the Pax Americana in the late 1970s and, consequently, the loss of American hegemony.[9] Thus in the aftermath of the 1990 war against Saddam Hussein, we offered the following conclusion:

> In the past [i.e., the era of Pax Americana], the United States had sufficient [epochal] hegemony to maintain the world [according to its historical mission.] Now, it is striving to maintain [the nostalgic impression of] that hegemony. Hence, we now see the type of military intervention that neither serves American interests nor substitutes for political [and economic] weaknesses. Thus, at this historical juncture, attempting hegemonic reassertion through aggression proves contradictory and self-limiting, thereby bringing to the open the most critical aspects of U.S. participation in [escalating] the present global disorder.[10]

It should not be lost to the reader that both the theoretical as well as policy implication of the context presented above is in variance with the majority of writers on this subject. For some, the question of war and peace pivoted around the alleged US need for more energy. Therefore, the whole conversation was decidedly changed to "To Need or Not to Need," perhaps in a comic rendition of Shakespeare's tragedy—Hamlet. Hence the peace or the war finds solution in an equally perverted option of US "energy independence." To the present writer this verdict is not only wrong in an ordinary scientific signification but also morally wrong for *reassigning* the cause and culpability of the war to a pointless popular *myth*, and participating in this campaign of misinformation (and disinformation), covering the criminality of conducting this war (and other wars) for the sole purpose of hegemonic reversal. As is explored further below, this is the most unforgivable facet of the slogan "No Blood for Oil," a facet that boarders on outright criminality.

## Oil and the Mismeasure of "Blood"[11]

As has been demonstrated in previous chapters, the history of Middle Eastern oil, including its subsequent development into a modern industry, has gone through distinct stages of development. The cartelization of oil in its all-inclusive and deliberate configuration was inaugurated with the "As Is Agreement" in 1928, which brought the entire oil business—from exploration, development, production through to transportation, refining, marketing, and retailing across the world (minus the Soviet bloc)—under the International Petroleum Cartel, known as the "Seven Sisters."[12] The cartel, however, began to lose its grip during the 1950s and 1960s. Proliferating market forces, in conjunction with the development of capitalist social relations in the colonial and semicolonial oil regions, had gradually overcome the longstanding colonial oil concessions, in the Middle East and elsewhere, brought in a collection of marginalized oil companies, known as "independents," and put an end to worldwide ironclad control of oil under the auspices of the International Petroleum Cartel (IPC) (1928–72).[13] As we have explicated earlier in this book, the crisis of 1973–74 was but the very symptom of this universal transformation toward globalization and the law of one *value* in oil. The crisis was an objective vehicle for a worldwide rationalization and restructuring of the entire oil sector on a fully capitalistic basis free of the archaic elements and insidious remnants of the so-called Postwar Petroleum Order—an order that is brandished as positive in Yergin (1991). Moreover, the so-called "OPEC offensive"—probably the most misconstrued part of oil history and thus unsurprisingly misperceived by the entire political spectrum, from the right to the left, as the cause of *recontrol* of oil—was but a catalyst concomitant with the decartelization of oil that promptly prepared the way for global competition, proliferation of the spot and futures markets, and globalization of oil.[14] As we have demonstrated earlier in this book, characterizing OPEC as monopoly or cartel shows an absolute lack of appreciation of oil in its modern connotation and, by implication, a lack of adequate awareness of the present-day global polity and politics, which inevitably intersect with it in the contemporary world.

As can be seen, the war-for-oil scenario is a popular myth that ignores the deeper understanding of the complex web of contradictions and the regulating dynamics of today's economy and polity. Yet, the very anachronism of this scenario is understandable in view of the archaic attitude and conduct by the US government. But all these atrocious acts are contrary to the authenticity of *global oil* that is so resistant to a devolutionary degradation that could possibly take us back to colonial times. As Karl Marx would have put it, this is clearly a matter of "social relation," not the subject of diktat by the United States. Yet, these atrocious acts are all explicable by their pernicious attachment to America's missing hegemony and the wishful attempt to get it back against the oceans of epoch-making waters that have already passed beneath the bridge of time, and a global polity that is so determined not to look back at the rearview mirror of America. Therefore, parallel with this anachronism, in light of the outrageous colonial comportments by the United States in Iraq, the shallow sham and shadowy *perception* of the "oil grab" has turned to a solid and sound "reality" for those who uncritically and unwisely chanted "No Blood for Oil." Yet, holding a parallel between the US invasion of Iraq and the control of oil is a fantastic proposition that was first intimated by deep-seated orthodoxy in mainstream economics and politics, long before liberals and leftists grabbed it. The persistent reality of oil since the mid-1970s decartelization pertaining to abolishment of the preconditions for the need of physical control and "access," was not heeded by either the conservatives, liberals, or, for that matter, radicals—including self-described Marxists. This rather counterintuitive reality, which also renders any connection between the war and oil superfluous, other than the claim of (given) oil rents by the invading forces, has been too hard to digest by these protagonists. On the orthodox economics side, commenting on the first US invasion of Iraq in 1990, James Schlesinger remarked:

> The United Stated [during the G. H. W. Bush administration] has gone to war now, and the American people presume this will lead to a secure oil supply. As a society, we have made a choice to *secure access* to oil by military means. The alternative is to become *independent* to a large degree of that *secure access.* (1991, emphasis added)[15]

At the time, it was astounding to the present writer as to why a distinguished Harvard economist, who had an illustrious carrier at the helm of the CIA, US Defense Department, and US Department of Energy, among others, failed to see the formation of (spot and futures) oil prices on a daily basis—notwithstanding, the interconnected, interdependent, and globally unified markets around the world—and still speaks like a character from outer space, alienated from the earthly reality of oil. As the brief passage above reveals, Schlesinger, stresses the phrase "secure access," on the one hand, and expresses desire for "independence" from the "secure access," on the other hand. There is a bit of subtlety in this twofold emphasis here. First, the so-called security of access to Middle East oil has no sense other than as a veiled substitution for the desirability of past history in which almost everything was made "secure" under the control of the IPC. Second, the desire "to become independent . . . of that secure access," is the obverse of this old neocolonial thesis, namely, an empty promise of achieving independence from Middle East oil in an interdependent world in which all oil turns into one undivided single pot. Thus, Schlesinger provides a convenient cover for two separate strategies: justifying the war without exposing its real cause, and creating panic, albeit indirectly, by playing the familiar scarcity card to extend the exploration of oil in the pristine US regions of wildlife, such as ANWR. This also explains the Bush-Cheney administration's attitude in getting the ball rolling toward the unregulated (and unmitigated) domestic oil exploration without an environmental protection oversight. In this regard, the Report of the National Energy Policy Development Group in Washington, known as Cheney's "Task Force on Energy," despite crafting and choreography by an assortment of professional lobbyists on hand, should speak for itself. In the report we read:

> The Alaska National Interest lands Conservation Act expanded ANWR from 9 million acres to 19 million acres, and designated 8 million acres as wilderness. Congress specifically left open the question of management of a 1.5-million-acre Arctic Coastal Plain area of ANWR because of the likelihood that it contains significant oil and gas resources... Section 1003 of the Act prohibits leasing of the 1002 Area until authorized by an act of Congress... In 1995, both the Senate and the House passed legislation containing a provision to authorize leasing in the 1002 Area, but the legislation was vetoed... The total quantity of recoverable oil within the entire assessment area is estimated to be between 5.7 and 16 billion barrels (95 percent and 5 percent range)... Peak production from ANWR could be between 1 and 1.3 million barrels a day. (White House (2001) *National Energy Policy*, May, Chapter 5: 9)

To appreciate the extent of fear mongering on energy by the Bush-Cheney administration, the report also insinuated that "ANWR production could equal 46 years of current oil imports from Iraq" (White House (2001) *National Energy Policy*, May, Chapter 5: 9). It should not be overlooked that Iraq (which had not yet been invaded at the time) was singled out off-the-cuff, in this report, as if the oil from any location in the world can be kept apart from the blended pool of global oil market. These lobbyists also tried to exaggerate projected flow of oil from ANWR, first, by padding their figures on the 5 percent probability of 16 billion barrels yield and, then, by using a "peak production" figure, which has nothing to do with average production capacity over the lifetime of the reservoir. This shows to what length the Bush-Cheney administration had gone to circumvent the obstacles that were initiated by environmentalists for years against domestic drilling of oil, particularly in pristine areas of wildlife in the United States. Besides, speaking of a need for Iraqi oil is a ruse, particularly that the United States itself was responsible for placing the harshest (UN) sanctions on Iraqi economy and society, including the import of medicine and export of Iraqi oil, which, among others, led to unnecessary loss of life of thousands of Iraqi children and to the decline of a vibrant Iraqi oil industry. Another point of critical comparison is the date of this report, that is, May of 2001, and the September 11, 2001, attack that devastated the twin towers in Wall Street, and the Pentagon. Why critical? Because it clearly shows that the invasion of Iraq had been contemplated long in advance without an inkling about the occurrence of the latter, perhaps at the inauguration of the Bush-Cheney administration in January 20, 2001.

Thus subterfuge of 9/11-reasonsing has a chance when Hell freezes over (see Eichenwald 2012 for details in this tragedy). The 9/11 incident was a godsend to the son, namely George W. Bush (and company), just as Saddam Hussein's invasion of Kuwait—albeit with a bit of help from April Glaspie, then US ambassador to Iraq—was a godsend to the father, namely, George H. W. Bush (and company), who engaged in a full-scale military mobilization across the entire region, and despite the final 48-hour diplomacy and Hussein's readiness to back off, to engage in "Operation Desert Storm" in the nick of time as the Soviet bloc was crumbling. What was an immediate accomplishment of the latter? Taking Bahrain apart and reassembling it as home of the US Fifth Fleet in the Persian Gulf, and strong-arming the UN General Assembly to revote and revoke the "Zionism is racism" declaration that had been passed rather overwhelmingly by the United Nations in the 1970s. But just as every father and son are in need of a ghost (holy or unholy) to accomplish the transformation, the latter revealed itself indubitably and fused the two in the free fall of American hegemony. On the issues surrounding the Israel lobby and US foreign policy, and "crisis of Zionism" see Mearsheimer and Walt (2007) and Beinart (2012), respectively. For the backdrop of the Israeli-Palestinian conflict and its complex and excruciating history Jabotinsky (1937), Brenner (1984), Pappe (2006), and Sand (2012) are among the vast number of must-read sources.

Returning to the main argument, in a nutshell, this thesis pays no heed to (1) the *mutuality* of the oil producers and oil consumer and the requisite of both sides in selling and purchasing in the competitive global oil market, (2) the interdependence of the oil regions in the present interdependent world, (3) the formation of the global price based on the cost of highest-cost (US) producer—not the cost of oil in each individual regions separately, and (4) the formation of oil rents (effected and secured by the relation of landed property and capital) according to the prevailing differential costs across the various oil regions in competition. Here, as has been demonstrated throughout this book, the dramatized "oil dependency" is but an empty phrase in the view of the interwoven fabric of all oil regions of the world in the course of oil crisis of the 1970s.[16]

On the left and liberal side of the political spectrum, not very many leftists understood the full implication of this speculative and flimsy thesis by the right-wing economists, such as Schlesinger, and their likeminded lobbyist in the US Congress and in the public media. Its popularity thus trumped its logic; the authority outdid the fact; and saddest of all, neither the liberals nor the radical left had had a theory of their own. Thus, the left-wing liberals and their radical counterparts, without a scintilla of critical study, adopted this threadbare right-wing charade and dressed it up in colorful left-wing garb and then hurriedly hit the street to protest the "war for oil"; and again these same wonderful liberals (and radicals) put on the very same masquerade, this time for the sake of saving the environment; the aim has been US "energy independence" through the lowering of domestic oil consumption by cutting the oil imports from the Middle East. The Middle East is a constant variable in this equation for the right and for the left. This demonstrates that the left is advertently or inadvertently playing a part in the justification of war in the context of the alleged US energy needs—which is morally suspect and by far more worrying. Both wings invoke "energy independence," the right pushes for limitless domestic exploration of oil; the left opts for energy conservation. The irony is that Michael Klare, a darling of the left, is one of the staunchest defenders of this right-wing thesis. He declares, "Two key concerns underlie the Administration's [the Bush-Cheney administration] thinking: First, the United States is becoming dangerously dependent on imported petroleum to meet its daily energy requirements, and second, Iraq possesses the world's largest reserves of untapped petroleum after Saudi Arabia."[17] Klare, however, takes this thesis and legitimizes it one step further in a panorama of neo-Malthusian scarcity:

> Global demand for many key materials is growing at an unsustainable rate. As the human population grows, societies require more of everything (food, water, energy, timber, minerals, fibers, and so on) to satisfy the basic material requirements of their individual members... Because

> the production and utilization of these products entails the consumption of vast amount of energy, minerals, and other materials, the global requirement for many basic commodities has consistently exceeded the rate of population growth. (Klare 2001:15–16)[18]

This same message has again been reiterated in Klare (2004). Yet, Klare—who is perplexed by the gravity of US involvement in Iraq—is "compelled . . . to conclude that petroleum is unique among the world's resources—that it has more potential than any of the other commodities to provoke major crises and conflicts in the years ahead" (2004: xiii). Again, for Klare (and for many on the left) the specificity of the *cause-and-effect* seems to have no bearing on this historically unique epochal conflict—and his fascination with oil is so intensive that he fails to realize the need for a specific and independent analytical proof.[19]

We contend that the war-for-oil scenario is at best a text without a context. At the methodological level, the oil scenario is a remarkable example of a *post hoc, ergo propter hoc* fallacy, supplanting the actual purpose of the US military intervention in Iraq.[20] Moreover, by neglecting the profundity of the last several decades of systematic global transformation, the protagonists on the left and on the right on this critical question both have adopted an impressionistic, irrational, and a fatalistic view of the US global role.[21] The left tends to capitalize on an ad hoc conception of hegemony and the functionalist pivot of US military might.[22] For Klare though the global conflict "is entirely the product of geology," which incidentally speaks volumes about his competence and his intellectual rigor.[23] The right, on the other hand, tended to rely on perceived "unipolarity" of the global polity in the post-Soviet euphoria and underscored absolute necessity of the American presence in every nook and cranny of the globe. This was the 1990s. The right-wing hysteria has now subsided in view of the punishing reality on the economic as well as political fronts and, more importantly, the devastating effects of the war and setbacks in both Iraq and Afghanistan, and the immense political and ideological repercussions that may yet materialize at the doorstep of America. Long after the glitter of the post-Soviet moment fizzled out, the right appears a bit defused; but it has yet to cope with a flurry of twenty-first century challenges that are here to spell out the loss of the American hegemony on a daily basis and that there will not be another chance for another American century.

There are others on the left who argue that this war may not have been for oil in the interest of entire US capitalism, but rather in the interest of a segment, namely, of "U.S. oil corporations." Hence, they propose that the cost of war[24] is tantamount to a huge subsidy given to the oil industry by the entire society. These antagonists, while admitting that their oil scenario straight-up is not tenable, try the backdoor to bring in a new version with a bizarre reasoning. This is a fictitious argument originating from the indefensible assumption of "direct access" and the necessity of "direct control"—an incommensurable stipulation with the reality of globalization of the oil (and globalization in general) in the 1970s. This argument is also silly, because it reduces the entire social and material interest of the capitalist class in the United States to the alleged interest of a tiny fraction. To put it in mathematical logic, this reduces the whole set to its fractional subset—an evident flaw in the logic. Besides, appealing to casual observation—such as viewing news from the Iraqi oil fields and the arrival of oil service contractors for "rebuilding" Iraq—this does not provide sufficient proof for a serious argument about the alleged importance of oil.[25]

Attaching significance to switching of foreign currency reserve holdings, from US dollar to euro, by OPEC oil producers is unjustified. As Paul Krugman pointed out in a short note, any possible shift from US dollar to euro by OPEC amounts to a "small change," for the US economy—much smaller than the switching made already by the "Russian Mafia."[26] This is a typical question owing to insufficient grasp of the post-Bretton Woods world in which the detachment of US dollar from gold, in 1971, had paved the way for emergence of other international currencies to flourish and in time to be designated as a source of global exchange. Moreover, the collapse of the Bretton Woods system is nothing short of pulling the rug out from underneath the *exceptionalism* of the US dollar rather forcefully signed in 1944 in Bretton Woods. The removal of this plank was already embedded in the development of European euro as a competing currency—that is, at its very inception. This amounts to a transmutation of the international monetary system. And this was not a *cause* but an *effect* of awesome and systemic changes, which were brewing beneath the exterior of Pax Americana since the late 1960s and which set off the beginning of the end. The question of switching currency is an epiphenomenon unworthy of discussion at this level of analysis, and by itself has no connotation in the context of a causal explanation. By way of an analogy, the so-called switching-the-currency as the cause of the US invasion of Iraq is not unlike the reasoning by certain economists, known as "circulationists," whereby the cause of economic crisis is to be found in the sphere of exchange (and circulation), rather than the sphere of production (see Weeks 1983). Again, the impression is the effect and the effect is the impression. However; many on the left are not losing any opportunity to grasp at this teeny straw.

Finally, it would be worthwhile to respond to a particular view that attributes the cause of "the Gulf Wars" to "securing property rights." The trouble is that this interpretation starts off with a reasonable expression of the relationship between capital and landed property in Marx as follows:

> For Marx, the particular form and content taken by landed property has a more-or-less *favorable* impact upon the accumulation of capital both as potential promoter of, and barrier to, access of capital to the land. Consequently, Marx's interminable numerical examples in Volume III of *Capital* are concerned to address how landed property can intervene in the process of accumulation (rather than as monopoly or differential rent depending on market or "natural" powers, respectively). (Fine 2006: 139, emphasis in original)

The above passage is written by one of the best and brightest economists in the world, and someone to whom the present writer owes an immense intellectual debt, and has contributed in a major way to Marxian value and rent theory. Yet Ben Fine tends to nullify the import of the above theoretical appraisal, which is true in its own right, according to the passage that follows. He speculates on the cause of US intervention in the Middle East by invoking the notion of landed property as a "barrier" to access of capital to land. He does this while knowing quite well that one would not know in advance the circumstances surrounding the access of capital to a particular land, given the divergences of forms of landed property. In other words, Fine knows very well that rent (in Marxian framework) is not a general theory but a specific theory that seizes upon the effects of a particular landed property. And this is what makes the valorization of landed property dependent on a detailed study of industry in question. Notwithstanding these caveats, we read:

> In this respect, the Gulf wars, and more, are about securing property rights to a highly land-dependent condition of production...As a consequence, securing property rights in the broadest sense, takes precedence even over (immediate) profitability. How otherwise can we explain the destructive effect of war on capital and profitability? Moreover, the means (and meanings) and mechanisms for securing property rights are necessarily distinct from the immediate accumulation of capital itself. (Fine 2006: 140)

Since the oil crisis of the early 1970s and the following decartelization and globalization of oil by the late 1970s, we do not find a scintilla of evidence that access by capital to the Middle Eastern oil deposits has encountered a barrier in the context deliberated above. Indeed, if anything, the opposite is true, given the fact that the majority of oil-producing governments in this region have often encouraged capital investments for further exploration, development, and thus adding to their oil reserves. Yet by invoking the question: "How otherwise can we explain the destructive effect of war on capital and profitability," Fine not only confounds the issues but also stretches the limits of economic determinism, perhaps inadvertently. Lastly, he is silent on pertinent issues surrounding the fall of American hegemony and US interventions that smack of returning to a bygone colonial era. Hence "securing the property rights" would obfuscate the real issues that are both theoretical and vital from the standpoint of policy and action in this area.

## Hegemony, Consent, and Mediation

The concept of hegemony, whether intended for examination of domestic class relations (à la Gramsci) or for investigation of international economic and political relations, must be understood in an "organic" and an indivisible context with reference to the constituent economic, political, and ideological aspects of the domestic society and global polity, respectively. The indivisibility is embedded in the very meaning of hegemony by virtue of the embedding interface of the economy, polity, and the society as parts of an organic whole. Therefore, to put it in methodological terms, hegemony is an *epistemology* informed of its *ontology*. In other words, a genuine *concept* shows bare unity of its subject, stripped of all encumbered divisions, fragmentations, and idiosyncrasies that may appear on the cloaked surface of the chaotic corpus. This is a good description of hegemony, a notion that is all about the fortitude of the whole—not the posture of parts in isolation. The whole also is the domain for interactions of the parts. Hence the interactions among the fragmented parts themselves are also beyond the fortitude of each individual part. That is why speaking of economic hegemony, as opposed to, say, political or military hegemony, is a ludicrous proposition. Likewise, it would be imprudent to assume that a power can have "military hegemony" but, not, say, "political hegemony." Such dichotomy is a cardinal sin in methodology and thus has no place in the *organic* domain of hegemony, which is rather analogous to the immune system in the biological world. It is in this context that hegemony would have a splendid application in the study of political power, power relations, and the shift of power in the global polity, which in turn could have relevance to the ascendance or the decline of a particular hegemonic power in a particular epoch.

The *internal* dynamics of *consent* and *force* are part and parcel of the intrinsic ideological power of the whole, which in turn provide the ingredient of leadership in a hegemonic system; this is far from any aggressive power projection by *external* means. This, in a broad measure, reflects Gramsci's hegemony and its application to the direction of national class struggle; this is also applicable to a broader arena of international economy and polity, particularly to the assessment of the rise and fall of the Pax Americana (1945–79). Consent and force are the necessary ingredients of the concept of hegemony. Yet, just as indivisibility of the time and space in modern cosmology, consent and force are not *separable* and have no consequence in the absence from each other. Moreover, parallel with the concept of "space-time," the signature of hegemony is "consent-force," representing a unified quality without fragmentation. Hence hegemony does not operate as a switchboard between a detached *consent* and a standalone *force* at the discretion of a supposed hegemon. Consent and force shed their usual linguistic connotations as soon as they find a synthetic unity within the concept of hegemony. Thus hegemony is essentially a synthesis that neither can be identified with consent nor with force in their standard linguistic meanings—much like the concept *space-time*. This will bring us full circle as to the real meaning of hegemony in terms of the indivisibility of economic, political, and ideological aspects of hegemony and the hegemon. Gramsci is rather explicit about antithetical coexistence of hegemony and the military force as follows:

> As soon as the dominant social group [i.e., the hegemon] has exhausted its function, the ideological bloc tends to crumble away; then "spontaneity" may be replaced by "constraint" in ever less disguised and indirect forms, culminating in outright police [i.e., military] measures. (Gramsci 1971: 60–61)

Similarly, the historical and institutional reasons for the erosion of American hegemony should be examined in conjunction with the mediating imperatives that have already generated a powerful tendency toward a multipolar polity and diffusion of power manifested in what is now identified as *globalization*. Gramsci, nevertheless, focuses on the "organic intellectuals" in modern society and then examines the formation of ideology and the "world of production" (i.e., the domain of purposeful human activity in capitalism) *mediated* through the complex intricacies of "civil society" and "political society" (Gramsci 1971: 12).[27]

In contrast, the pervasive notion of hegemony popularized in the orthodox International Relations literature refers to *unmediated* impositions that more often than not capitalize on the direct antagonistic interactions among the opposing counterparts in the polity, particularly the international polity.[28] Simply put, "hegemony" so defined is not significantly different from *domination*. Indeed, a remarkable journalistic replication of such an interpretation is so widespread in the media that, despite the lack of critical justification, it is nevertheless appealing to popular culture everywhere across the global divide. Therefore, it is often painstakingly difficult to get through an intelligent discussion without a great deal of digression and considerable shift of focus when, for instance, it comes to the question of the hegemonic rise and the demise of Pax Americana. Hegemony, so perceived, is an ad hoc, ahistorical, tautological, and an impressionistic buzzword employed not only by the majority of the public but also by many specialists across all social sciences—from the very left to the very right. Yet, the fault of the leftists is more pronounced than that of the right, given that the former have had more exposure to this term and its left-leaning origins. They clearly should have known a good deal more. The common understanding of hegemonic position in popular parlance is the imposition of force from without toward either annihilation or domination of one's enemy, rather than relying on the mediating institutions toward a hegemonic solution. For instance, the US involvements in all three Persian Gulf Wars (including *proxy* war of 1980–88 between Iran and Iraq) were falsely branded as hegemonic.[29]

Therefore, force and domination by military means should not only be deemed as the negation of hegemony but a quintessential measure of hegemonic decline. To put it simply and bluntly, one does not burn the house in which one has shared ownership and hegemonic interest—even the Cosa Nostra knows this. A tangible example of hegemony can be seen in the case of employment (of wage labor in all its manual and mental manifolds) in capitalism. Given the everyday struggles of the majority of people in the world to make ends meet—and entirely aside from Marxian theory of employment—there is little doubt that working is not without exploitation. But against this horrendous feeling, people are looking for a job, any job, all the time. Indeed, people beg to have a job; they beg to be so exploited. This is a remarkable illustration of omnipresence in capitalism. And it is the very stuff of which hegemony is made. Consent is thus embedded deep in the structure of hegemony; and it would be silly to think that it can be manufactured in a rough-and-ready, strong-armed fashion.

Hegemony thus must have several interwoven characteristics as follows. It should be (1) spontaneously consensual, (2) internally determined, (3) historically sanctioned, (4) holistically undivided, (5) structurally synthetic, and (6) institutionally mediating. These characteristics are inseparable and must be present for hegemony to take hold. In other words, these are the vital signs of hegemony as opposed to, say, domination, partisan control, and forced invasion. The story of hegemony is the mutuality of the whole, albeit with its own internal contradictions and far from romanticized view of harmony in orthodox terms. Our focus here is particularly on the rise and the fall of hegemonic Pax Americana, a historically specific interstate system that had risen after the Second World War (1945) and fallen in the late 1970s. Hegemony is a mutual characteristic of the system as a whole—not a separate property of the hegemon. American hegemony, therefore, should not be seen as one-sided and in mechanical terms but dialectically.[30] Hegemony thus thrives through reflection of the *whole*, not exertion of the parts. To be organic, the (subjective) hegemonic power of a "social group," "a national entity," etc., presupposes the material, social, and historical conditions through the objectivity of the whole. Otherwise, any claim to hegemony dwindles to an algebraic sum of its constituent parts and inevitably reduces to a *fallacy of composition*. For that reason, the United States had gained hegemonic power by the virtue of the system of the Pax Americana as a whole, not by the magic of the inherent grace credited to American ingenuity or worse—American exceptionalism. Thus, in the absence of hegemonic Pax Americana, speaking of American hegemony has no consequence or meaning whatsoever.[31]

## The Rise of the Pax Americana

To appreciate the concrete manifestation of hegemony in the then ascendant Pax Americana[32] we simply need to focus on the application of the (tripartite) American Doctrine of Containment after the Second World War. This doctrine was an embodiment of (1) the containment of the Soviet Union, (2) the suppression of democratic/nationalist movements across the "Third World," and (3) the control, co-option, and constriction of domestic "civil society" within the United States.[33] This three-prong strategy is neither adequately theorized in its interlace within the orthodox international relations literature nor fully categorized in the heterodox counterpart. From the standpoint of hegemony and despite their causal juxtaposition by mainstream theory, these three containments should be treated as an articulation of the organic whole in the newly formed postwar international polity.

The first containment crystalizes the placing of the Soviet Union behind the ideological divider of the Iron Curtain and the initiation and the material enforcement of the Cold War from the beginning.[34] Hence, the Cold War was indeed a manifold hegemonic phenomenon, spanning the economy, polity, and the realm of ideology and culture, from the race for space, the research and development, the proliferation of consumer goods, the fetish of openness, to Hollywood, Mickey Mouse, rock 'n' roll, and hula hoop. Evidence for the second containment can be deciphered from the formal declaration of anticolonial policy, on the one hand, and subversion of nearly all democratic (national) movements across the "Third World," on the other hand. This doctrine, as has been demonstrated earlier in the removal of Mossadegh in Iran, often led to covert campaigns and coup d'états mediated through the handpicked regimes which, despite their contradictory outlook and archaic political attitude, were nevertheless an embodiment of Pax Americana itself.[35] At the same time, US attempt at ad hoc and lopsided economic alteration of these regions—for instance, via the cookie-cutter land-reform programs—led to their hasty, paradoxical, and undemocratic inclusion in the transnational capitalist orbit.[36]

Finally, the third containment strategy was channeled through American domestic thought control and the deliberate expulsion of the independent institutions and organizations that had been a part of US domestic "civil society"— such as militant labor unions and the progressive political institutions of the real or imaginary left. This, in turn, had given rise to fashioning a "hegemonic model" for broad ideological emulation that predictably shifted the direction and scope of American popular politics to the far right of political spectrum for good, and marginalized the left. McCarthyism is but the tip of this iceberg when one comes to the assessment of this strategy and its insidious effects on American psyche. As we have maintained elsewhere, "America itself has become a prime casualty [of this doctrine] that has yet to recover from its prolonged [intellectual] timidity, its chronic insecurity, and above all, the injuries sustained to its collective consciousness."[37]

In the economic arena, the indication of hegemony can be seen from the status of the US dollar as universal currency in connection with the newly devised international monetary system, known as the Bretton Woods (1944–71). This arrangement—coupled with the Marshall Plan for the postwar reconstruction of Europe and Agency for International Development (AID) for the "Third World"—exponentially increased US hegemonic power during the formative years of Pax Americana. Institutionally, powerful US-sanctioned international institutions, such as the International Monetary Fund (IMF) and the World Bank, came to direct and influence the modus operandi of particular development strategies that paved the way for hegemonic ascent of the system as a whole. Indeed, the socioeconomic restructuring of the Pax Americana was preconditioned upon the eventual *absorption* of nearly two-thirds of earth's inhabitants across the vast geography of production and exchange, known as the Third World. Such absorption was, of course, contingent upon what Gramsci calls "passive revolution," in which the separation of an immediate producer from the land creates an internal market for propagation of the capitalist accumulation and penetration of transnational capital.[38] This is contrary to the colonial system of the Pax Britannica that essentially lived and died by the sword through (1) *direct* colonial administration, (2) *direct* plunder of natural resources, (3) *direct* and unmediated repatriation of surplus (and wealth) from the colonies, (4) foreign policy based upon the universal application of *gunboat* diplomacy, and (5) unrefined and explicit institutional racism, physical segregation, and cultural *separation* within the empire, far from *mediating* institutions conducive to integral development. These characteristics carry no hint of hegemony more than a bully across the schoolyard who traumatizes weaker and vulnerable pupils to get something for nothing. Therefore, anyone who would assign hegemony to the British Empire, with respect to international relations, could not be more wrong.

Now, given the depiction of Pax Americana above, the extent of hegemony embedded in the system can be measured by the accommodating socioeconomic forces that assumed the role as foundational for this international order at the time. This is a necessary condition for realization of hegemony. On the other hand, the postwar system of Pax Americana enjoyed the mediating economic, political, and ideological institutions that proved compatible with this foundation. The latter are sufficient conditions for hegemony to take hold. Both of these categories (of conditions) must be satisfied to speak of hegemony in a modern socioeconomic system. In the case of the Pax

Americana—during its height and hegemony—social relations, combined with international economic institutions, led to the evolution of capital in its full-fledged transnational form. American hegemony so acquired was by no means secure. From time to time, it suffered from contradictory (either systemic or discretionary) challenges that placed it in jeopardy and brought its fragility in the open. The conduct and aftereffects of a series of CIA coup d'état are testimonial to this hegemonic fragility.

To be sure these and other political and ideological atrocities perpetrated by the United States are unbecoming of hegemony and thus showed the open-endedness and instability of US leadership in its different moments. All these coups express a state of suspended animation between a "war of movement" and a "war of position" in Gramsci's parlance (see Bocock 1986).[39] Yet, the United States survived all this only to be taken down by the impact of social relations themselves. It was the evolution of social relations and thus globalization capital—notwithstanding the global consequences for new economic, political, and institutional requirements—that eventually unraveled the Pax Americana; hence the newly developed and transmuted body (i.e., the sum total of global economy, polity, and society) within had exceeded the epochal capacity and historical necessity of the old vessel. In other words, the contained contravened the container. This, by implication, speaks loud and clear on the closure of American hegemony and the end of US leadership as a global hegemon. Indeed, there will not be any such hegemon in the present epoch where the "conquest of the mode of production" across the world is realized and the capitalist social relations are in the seat of global hegemony.

## The Demise and Puzzle of Hegemony

As we have seen, the overriding reason for the implosion of Pax Americana was a series of cumulative changes that eventually led to an epochal qualitative change within the global socioeconomic relations in the 1970s. The enormity of this grand sea change also displaced the constituent parts of the system, particularly the hegemonic position of the United States at the apex of the "old" international system. Therefore, the claim of "unipolarity" (of the emerging global polity) made by many in the international relations circles, and elsewhere, is a superficial claim at best, and deceitful at worst. At a more fundamental level, globalization is the mirror of mobility of capital across the boundaries of nation-states, which led to devastation of the preexisting national hierarchies, international division of labor, time-honored locations of technology and technological development, and boundaries of markets, among others since the 1970s.

The so-called *plant closings* across the United States is the story of a grand sweep of the rationalization of manufacturing and other related industries across-the-board and across the global divide in swiftness and velocity unparalleled in recent memory. The so-called outsourcing cut across-the-board to the bone the industrial base of once prosperous postwar America. This was happening in the 1980s when Ronald Reagan was riding high and on the surface all seemed quiet on the Western front. But in our estimation it was, in an objective sense, a mop-up operation that spelled out the beginning of the end of American hegemony (see Bina, Clements, and Davis 1996). And all this was through hyper-competition and the fast-paced technological change across the worldwide geography of production. In political and institutional terms, the European unification and political and ideological upheavals that, since the late 1970s, have taken place in many strategically located client-states within the Pax Americana clearly speak to this epochal change.

It is worth mentioning that the rise of the so-called political Islam, since the late 1970s, should not be interpreted as an independently competing sociopolitical alternative to liberal democracy or socialism, despite the convoluted rhetoric of Islamists and Islamic movements, and in spite of terrifying, prejudicial, and embellished claims in the Western media. But it would be a mistake to think that these movements or the impetus given to their rise are ad hoc or unrelated occurrences in this post-Pax Americana world. Many of these movements are the epitome of repression linked to the US containment strategy in the "Third World"—indeed the chickens of the ancient regimes have now come to roost. But, as the Iranian experience demonstrates, Islam has not proven a viable third alternative on its own footing, either to liberalism or socialism, or a sustainable substitute for the plural civil societies of the post-Pax Americana. Yet, the rise of political Islam presents the mirror image of the political failure—and indeed the demise—of the West and Western liberalism. Given the collapse of Soviet "state capitalism" and the crises of Western liberalism, neither of these ideological models appears viable in the eyes of alienated Islamists. Nonetheless, as the byproduct of the very system itself, the rise of political Islam, for a good measure, reveals the loss of American hegemony.[40]

Now, further examination of the desperate and discursive reversal of these and many other aspects of US hegemony, particularly since the beginning of the Bush-Cheney administration in January 2001, reveals an absolute absence of American hegemony in terms of the complete replacement of spontaneity with belligerence, multilateralism with unilateralism, and containment with a reckless preemptive strategy. In December 2001, the Bush administration unveiled its "National Strategy to Combat Weapons of Mass Destruction." This was following its earlier pronouncement of the axis-of-evil policy against a handful of so-called rogue states in the Middle East and elsewhere. The Bush administration used the unfortunate events of September 11, 2001, as a convenient cover to inaugurate its new doctrine of permanent war.[41] This new Doctrine of Preemption is as follows:

> An effective strategy for countering WMD [Weapons of Mass Destruction], including their use and further proliferation, is an integral component of the National Security Strategy of the United States of America. As with the war on terrorism [i.e., invasion of Afghanistan, etc.], our strategy for homeland security, and *our new concept of deterrence, the U.S. approach to combat* WMD *represents a fundamental change from the past*...(p. 1, emphasis added). Because deterrence may not succeed, and because of the potentially devastating consequences of WMD use against our forces and civilian population, U.S. military forces and appropriate civilian agencies must have the capability to defend against WMD-armed adversaries, including in appropriate cases *through preemptive measures.* This requires capabilities to *detect and destroy an adversary's* WMD *assets before these weapons are used.* (p. 3, emphasis added)[42]

As we have seen, following the full-scale invasion of Iraq by the American and British forces, no weapons of mass destruction (WMD) have been found. However, in the 1980s, while Saddam Hussein was using chemical weapons against the Iranian troops and Iraq's own Kurds, the US government, with full knowledge of these activities, showed no apparent concern about the use of the WMD.[43] The following passages are illuminating:

> The U.S., which followed developments in the Iran-Iraq war with extraordinary intensity, had intelligence confirming Iran's accusations, and describing Iraq's "almost daily" use of chemical weapons, concurrent with its policy review and decision to support Iraq in the war. The intelligence indicated that Iraq used chemical weapons against Iranian forces, and according to a November 1983 memo, against "Kurdish insurgents" as well.[44]
>
> [When] Donald Rumsfeld...was dispatched to the Middle East as a presidential envoy...[h]is December 1983 tour of regional capitals

> included Baghdad, where he was to establish "direct contact between an envoy of President Reagan and President Saddam Hussein," while emphasizing "his close relationship" with the president. Rumsfeld met with Saddam, and the two discussed regional issues of mutual interest, shared enmity toward Iran and Syria, and the U.S.'s effort to find alternative routes to transport Iraq's oil; its facilities in the Persian Gulf had been shot down by Iran, and Iran's ally, Syria, had cut off a pipeline that transported Iraqi oil through the territory. *Rumsfeld made no reference to chemical weapons, according to detailed notes on the meeting.*[45]

On the other side, some 20 years and two Persian Gulf Wars later, US officials—following intense pressure, disdainful bribery, and frantic illegal wire-tapping at the UN Security Council—proved unable to get what they asked for. This time, contrary to the Iran-Iraq war interlude, the Bush-Cheney administration insisted that Saddam Hussein must have been developing WMD. This claim, however, was incompatible with the various UN inspection reports under Blix and El-Baradei. Blix also challenged the United States for being unreasonable and, perhaps, unserious about the adequacy of time for thorough UN inspection.[46] The heavyweight members of the Security Council, aside from the United States and the United Kingdom, did not want to be rushed into premature action. They all preferred a peaceful resolution to a warring option at that time. It was the United States, which was anxiously looking at its preemptive watch—a bloody timetable that was set by the Pentagon group under Wolfowitz. Despite many opportunities for inspection, no WMD have ever been found in occupied Iraq.[47]

In the meantime, the UN Security Council presentation by Secretary of State Colin Powell turned out to be a hoax. It was a clumsy blend of outright plagiarism and deliberate misinformation. The so-called evidence was carbon-copied from three different sources, including a decade-old information reported in a graduate student thesis, portrayed as new intelligence.[48] He also made untruthful claims based on the alleged recorded conversation of two senior [Iraqi] officers. In his presentation, Powell also named an Iraqi business enterprise, which was allegedly involved in prohibited weapons system activity.

Finally, in a haphazard manner, Collin Powell, the Secretary of State in the Bush administration—(while George Tenet, CIA director, was sitting right behind him and leaning on his back)—showed the members of the UN Security Council several satellite pictures from the alleged sites that, in his own words, were the center for the production of WMD. At the end, Powell, who clearly realized that there was not much evidence on the table, was quick to declare, "I cannot tell you everything that we know. But what I can share with you, when combined with what all of us have learned over the years, is deeply troubling."[49] After a 16-month investigation and thorough examination of more than 40 million pages of documents, the Duelfer Report had found no evidence of WMD in Iraq.[50]

In sum, when the so-called good guy of this administration takes part in deliberate deceit and has no qualms with signing off and participating in a rogue, rotten, and smelly foreign policy, we need to look deeper into not only the temporal decadence of the Bush-Cheney administration but also the self-defeating effects of epochal loss of hegemony by a declining power, namely, the United States, that is now in a free fall. To be sure, 9/11 was the trigger, not the cause.[51] For the cause has already been embedded in the crumbling structure since the collapse of Pax Americana and the loss of US global hegemony. Contrary to the evidence, there are individuals who had a hand in all this in the Bush administration, and who are still insinuating on the alleged connection between Saddam Hussein's government and al-Qaeda. That is how the so-called fight against "terrorism" has gone through Baghdad. That is why a declining power is far more dangerous to world peace than a rising one.

## Concluding Remarks

History has proven that capitalism is not about self-sufficiency, security, and independence, much less about energy and oil independence. It is rather about discursive mutuality and contradictory interdependence. The war-for-oil scenario obtains its lineage from an old, speculative, and ahistorical right-wing economic theory where the right relies on its anachronistic application of oil monopoly and the theory-less and clue-less left on its romanticized interpretation. The oil, however, is the effect—not the cause—of the US invasion of Iraq. The search for the cause, however, should converge on the fall of American hegemony, which in turn is progeny of the collapse of Pax Americana. The loss of hegemony and attempting to reverse it, in turn, provoked the United States to engage in belligerent actions of monumental consequence. This propelled the United States even further toward the abyss of hegemonic descent. The conundrum of hegemony turned the United States into a self-flagellating monster that did not only engage in the misdeeds of war against humanity by slaughtering tens of thousands of innocent Iraqis, a staggering scale, but also committed the very irrational misdeeds against its own self-interest by hastening its further decline. This is irrationality of the first order, the kind of irrationality that is a million times more potent against world peace and stability than the atrocities perpetrated by some barefooted, rag-tag "terrorists" who have no stomach for secularization or foreign intervention. And it is a remarkable demonstration of *real* "war on terror" collapsed onto and on its own crumbling self.

Hegemony is a deliberate concurrence of consent and force in a historically specific and politically explicit setting in which spontaneity is the order. Consent and force are not separate entities but a melded part of a synthetic whole, thus negating any reduction to either of its constituent parts. Therefore, force, particularly dominating force, has little to do with hegemony. At the same time, such synthesis does not lend to indivisibility of the economy, polity, and the social relations. In other words, hegemony represents all these aspects in an organic unity. Thus military superiority is not a sign of hegemony. Indeed, the popularized phrase "military hegemony" is a contradiction of terms. In the light of all this, it would be appropriate to say a few words in passing on the "inevitability" of US global leadership as a *bound to lead* power and the notion of "soft power" in two notable books with same consecutive titles (see Nye 1991 and Nye 2004). Soft power seems to suggest power by persuasion (and co-option), and in a way tends to mimic "spontaneity," which is one of the differentiating aspects of hegemony. But global hegemony is not a matter of choice and thus cannot be obtained by mere persuasion, co-option, sugarcoating, cunning, fictitious "partnership," joint military exercise, "special relationship," "preferred trade," and, in a nutshell, by deliberate willpower—hegemony is not voluntary. Hegemony is (historically) unique, just like virginity; when you lose it, you lose it forever. This has little to do with intention and a lot to do with the obtainability of the objective conditions that would delineate and pair off a period and a polity in unison. These propositions sum up to icing on the cake where there is no cake—a pretense of leadership. Hence the United States is neither bound to lead nor bound to be a hegemonic power, even in mimicking rendition of the "soft power." So, the thesis of "soft power" by Nye seems a bit like placing the cart before the horse, so to speak, by putting the hegemony ("soft power")—the consequence—before hegemony ("soft power")—the premise—in the tumultuous post-American global polity in the making.

The Bush-Cheney administration itself was a temporal sideshow, a sticky tangent to the vortex of an American downward spiral. The

Administration's three-prong political base, namely, the neoconservatives, Cold Warriors, and Christian (Zionist) fundamentalists was the main show (within this sideshow), which has done the highest bidding for furthering US decline. There is also a detectable tendency toward a social rupture on the domestic side of the United States, which is very terrifying.[52] The significance of oil had been limited by the trickle of Iraqi oil production under siege by the insurgency. This was, nevertheless, a godsend for the disbursement and fattening of the likes of Cheney's Halliburton. Finally, as the US combat troops were beginning to withdraw during the Obama administration, the time for signing of the oil contracts had arrived. And, as the present writer had envisaged long before the war, the US companies were not the sole winners in this global game. In the most concrete level of observation the question of oil turned moot. But even so, there are many who still depend on their eyes, as Mark Twain suitably expressed, when their "imagination is out of focus."

The world is now grappling with the loss of American hegemony and the debilitating aftereffects of reckless US reactions. The epochal train had already departed from the good old station, and the passenger who is running wild in the opposite direction through the rear cars is running out of time.

# Notes

## Introduction

1. F. Fukuyama (1992), *The End of History and the Last Man* (New York: Avon Books).
2. T. Shanker and E. Schmitt (2004), "Pentagon Weighs Use of Deception in a Broad Arena," *New York Times*, December 13: 1, 12.
3. It is instructive to revisit the documents in this period on the issue of nationalization. For instance, being in cahoots with Nixon Doctrine in the Persian Gulf, Saudi Arabia all along did not wish to nationalize its oil and was asking for better "equity participation" with the International Petroleum Cartel (IPC). Saudi officials (including Sheikh Ahmed Zaki Yamani, the petroleum minister) asked for gradual increase in their "share" before they opted for 100 percent equity. This is but a de facto nationalization of oil. The title of following piece, however, speaks rather vociferously on political timidity of Saudi Arabia and the extent of regime's captive status by the United States. See A. Z. Yamani (1969), "Participation Versus Nationalization—A Better Means to Survive," *Middle East Economic Survey* 12 (June 13).
4. C. Bina (2012b), "Who's Afraid of Kirchner's Oil Nationalization," *Asia Times*, May 8.
5. C. Bina (1985), *The Economics of the Oil Crisis* (London and New York: Merlin and St. Martin's); C. Bina (1990), "Limits to OPEC Pricing: OPEC Profits and the Nature of Global Oil Accumulation," *OPEC Review* 14 (1): 55–73; C. Bina (2006), "The Globalization of Oil: A Prelude to a Critical Political Economy," *International Journal of Political Economy* 35 (2): 4–34; C. Bina and M. Vo (2007), "OPEC in the Epoch of Globalization: An Event Study of Global Oil Prices," *Global Economy Journal* 7 (1): 1–49.
6. C. Bina and C. Davis (2008), "Contingent Labor and Omnipotent Capital: The Open Secret of Political Economy," *Political Economy Quarterly* 4 (15): 166–211.
7. S. D. Krasner (1978), *Defining the National Interest: Raw Materials Investments and U.S. Foreign Policy* (Princeton, NJ: Princeton University Press); D. W. Drezner (2011), *Theories of International Politics and Zombies* (Princeton, NJ: Princeton University Press).

8. See S. Bromley (2005), "The United States and the Control of World Oil," *Government and Opposition* 40 (2), Spring; S. Bromley (1991), *American Hegemony and World Oil* (University Park, PA: Pennsylvania State University Press); C. Bina (1994c), "Review of American Hegemony and World Oil," *Harvard Middle Eastern and Islamic Review* 1 (2): 194–98.
9. D. Stokes and S. Raphael (2010), *Global Energy Security and American Hegemony* (Baltimore, MD: John Hopkins University Press).

## 1 World Oil and the Crisis of Globalization

1. There is a growing literature that represents the main characteristic of this category. The results of a symposium held after the oil crisis, with contributors such as Raymond Vernon, Romano Prodi, Alberto Clô, Edith Penrose, George Lenczowski, and James McKie were edited by Vernon and published in *Daedalus* Fall 1975b. See also M. A. Adelman, 1972, and Taki Rifai, 1974.
2. There is an extensive literature on dependency theory, but perhaps the popular arguments can be found in Arghiri Emmanuel, *Unequal Exchange,* 1972 and in Samir Amin, *Accumulation on a World Scale,* 1974. See also Andre Gunder Frank, 1969a, 1969b, 1972.
3. Norman Girvan uses the phrase - "OPEC offensive" to show that the oil crisis was the consequence of the nationalism of the Third World.
4. For the analysis of monopoly and competition see Bina, 1985, Ch. 5.
5. It is important to note that here we are concerned with the average oil recovery per well, in association with the entire volume of capital investments that were applied to these oil fields, rather than what is known as the "investment at the margin," a prevalent concept in the neoclassical economics. The lumpiness of these capital expenditures and their role as regulator of the entire production aside from the quality of the oil deposits. This is significant for the determination of value based upon the least productive oil fields. For more discussion see Ben Fine 1979, Bina 1985, Ch. 5.
6. In the mainstream economic theory there are only ghosts of economic crisis rising from the dead and visiting upon the economy from time to time. There is little articulation and almost no theory as to the restructuring of the circuit of capital in micro (the industry level) or for that matter in macroeconomics (the economy as a whole). In Marxian theory, breakdown in the circuit of capital is anticipated within the structure of value theory to correspond with the reality of the capitalist economy—far from speculation. In the 1973–74 crisis of decartelization (i.e., competitive globalization) of oil, one must appreciate the objective side of this rather universal restructuring and its regulating mechanism within a broader context beyond the Middle East, and beyond the superficial interactions of the actors— irrespective of their political and ideological

intention. These ad hoc contingencies are worthy of consideration. But they do not provide us with anything tangible unless we have a theory that is beyond the actions of these and other agents to be able to articulate them. This is the difference between, say, astrology and astronomy in modern Physics; and it is certainly the difference between mainstream economics and its classical and Marxian counterpart.

## 2 WORLD OF MODERN PETROLEUM AND THE OIL RENT

1. Harvey (2003) stubbornly clings to the duality of "territory" (i.e., geography) and "capital" (i.e., social relation), particularly where it comes to the oil sector and oil fields. He then alleges that "OPEC monopoly," and recent US interventions in Iraq and Afghanistan are connected to oil and that the world has now entered an age of "The New Imperialism."
2. It is perplexing to me that the work of James Anderson (1739–1808) had been left without an acknowledgment by Adam Smith who was his contemporary and lived in close proximity in Scotland at the time. I have come across a copy of the first edition of Anderson's *Essays Relating to Agriculture and Rural Affairs* published in 1797 while searching for material in the University of Maryland library, at College Park, in 1974. I was astonished to know that Anderson had produced a more coherent theory of agricultural rent than Ricardo did. While I do not understand the hesitation of Smith, who might have come across this important source, in acknowledging Anderson's contribution to classical rent theory, I wish to bring this to the attention of scholars and students who may want to do further work in this field.
3. In neoclassical price theory "scarcity rent" is the measure of opportunity cost of intertemporal allocation. Hence, MC + Scarcity Rent = User Cost. This is the highest stage of initiation in the hall of fame in orthodox theory. Yet certain (post-Keynesian) economist has managed to cling to such a tautological ploy and at the same time hold on to heterodoxy (see Davidson 1979).
4. One of the valuable contributions of Ricardo is the distinction between profit and rent. Chevalier's framework negates this contribution and opens the door for amalgamation of capital and landed property. This, in a methodological sense, has divorced Chevalier (1976) from the classical school and wedded him to neoclassical economics, where anything could be rent, thus nothing is rent.

## 3 OPEC: BEYOND POLITICAL BATTERING AND ECONOMIC ROMANTICISM

1. Today, setting up competition and monopoly in their axiomatic and idealized forms in the opposite ends of the "market-structure" spectrum,

and attempting to bridge the gap by negation of *real* competition through the method of *successive approximation*, is but the basic tenet of the industrial organization literature.

2. Andrew E. Kramer, *New York Times*, June 14, 2012: http://www.nytimes.com/2012/06/15/business/global/opec-is-said-to-leave-production-steady.html.
3. For Schumpeterian theorists, competition is not an idealized rendition of perfect atomistic markets to be counterposed to equally idealized "monopoly" or "imperfectly competitive" markets in a tautological manner. Rather, competition itself is internal to the process of "creative destruction," and dynamics of concentration and centralization of production. Hence, the presence of integration does not necessarily negate competition. This is also generally true for the Classical, Marxian, and Austrian theories of competition (Schumpeter 1928, 1942, Ch. 7; McNulty 1967; Clifton 1977; Shaikh 1980, 1982; Weeks 1981a, Ch. 6; Semmler 1984; Bina 1985, Ch. 6, 1989a, 2006; Kirzner 1987; Bina and Davis 1996).
4. For critical examination of oil (and energy) as a decartelized global sector see Bina (1985, 1988, 1989a, 1992, 1994b, 1997, 2006).
5. Given the eventual physical exhaustibility of oil and thus intertemporal opportunity cost of its present and future exploitation, neoclassical theory contends that some measure of "scarcity rent" must be added to the marginal cost of oil extraction. Hence, MC + Scarcity Rent = User Cost. Adelman (1986, 1990) believes that oil is produced at the time of discovery; thus, reserves must be treated as "inventory." In addition, Adelman ignores the very distinction between the ownership of oil reserves and holding of such reserves through leases, which in turn separates oil rents from profits. Bina (1985) does not rely on the physical exhaustibility of oil but contends that the economic institution of modern landed property and the ownership of subsurface oil deposits are behind the distinction of categories of *rents* and *profits* in the oil industry. Hence, given the formation of competitive (general) rate of profit, differential productivities of the existing oil fields turn into differential oil rents across the globe. Theoretically, given the legendary debates within the classical school, these rents, while price-determined, not price determining, are still dependent upon the property rights of subsurface ownership. Thus, according to Bina, given the oil industry's high capital intensity, differential oil rents both for OPEC and non-OPEC producers are neither an arbitrary sum nor monopoly rent, and OPEC should not be seen as a monopoly or a cartel. According to this framework, it is also inaccurate to call such rents "Ricardian," since they not only automatically correspond with the differential fertility of land (i.e., natural objects) but also capture the effects of the institution of ownership in conjunction with such differential fertilities (see Fine 1979, 1983; Bina 1989b, 1990, 1992, 2006). Besides, in addition to the

rendition of Ricardo's theory of ground rent, the textbook (neoclassical) rendition of Ricardo's "comparative cost" theory also mischaracterizes Ricardo's combined system of international trade and finance.

6. Today in the global production of oil, differential oil rents are a permanent economic category whose existence does not wash away with "cheating," lack of coordination, and overproduction by OPEC and/or non-OPEC producers alike. Moreover, the magnitudes of these rents are not arbitrary, or subject to bargaining based on games theoretic consideration among members, but are established through intraregional and interregional competition. In addition, these rents should not be mistaken as "taxes," just because they are being obtained by the state. Thus, any contentious interactions, such as "bargaining" among OPEC members or between OPEC and non-OPEC producers, must be analyzed within the confines of such magnitudes (see Bina 1985, Ch. 5, 1989b, 1990, 1992, 2006). Consequently, it would not be sufficient to resort to "games theory," as the latter is belonging to the domain of apparent actions and reactions, and thus emerging effects, rather than being a suitable territory for searching for genuine cause.
7. Given the peculiarities of the petroleum industry and the specificity of its economic institutions, and contrary to the conventional arguments—that rely on cost accounting of oil reserve alone (see Adelman and Shahi 1989; Adelman 1992)—oil costs must include differential oil rents.
8. To date it would be difficult to find an adequate and judicious review of literature on oil. To our knowledge, Mabro (1998) presents the most recent review. Yet the author's constricted and unsystematic framework leaves hardly any room for recognition, and thus references to literature on differential oil rents, beyond his own narrow and habitual understanding. Mabro writes, "Our purposes, however, are much wider: to inform readers about the publications that are *worthy of attention,* to assess their major contribution and to present in conclusion a brief *personal interpretation* of the nature and role of OPEC" (p. 4, emphasis added). Another reviewer, an economist with steep neoclassical economic conviction and flat out ignorant of alternative literature on economic theory and oil, pretends: "In 1991 . . . I published a book that took *stock of the knowledge thus accumulated.* This article brings up to date the findings of that earlier survey of the economic literature on models of the oil market [*sic*]." (Salehi-Isfahani, 1995: 3, emphasis added.) Therefore, it seems that the so-called *stock* of knowledge in this case is, replicated once again, in *the flow* of ignorance, if not in the sentiment of deliberate exclusion. This attitude, combined with dogmatic attachment to the cul-de-sac of neoclassical orthodoxy appears to have been an essential ingredient for producing fake and fictitious analysis, and a recipe for frightful political judgment, unrelated to oil in today's world economy.

## 4 The Globalization of Oil

1. The Redline Agreement is the infamous Cartel decision, which made the Iraq Petroleum Company (IPC) conspire against Iraq and withheld 99.5 percent of Iraqi territory from any attempt at exploration. This arrangement was a part of a larger gentlemen's secret agreement made in the Achnacarry Castle, Scotland, in September 1928. For further suppression of oil discovery in the Middle East see Blair 1976: 81–85. Blair rightly observes:

   *Contrary to the widespread and long-standing impression that cartels are somehow inherent in the nature of things, the fashioning and implementing of these arrangements was the product of a great deal of very hard work.* According to the cartel's minutes, which came into the hands of the Swedish investigating committee, the group held 55 meetings in 1937 at which 897 subjects were discussed; in 1938, 49 meetings were held at which 656 subjects were discussed; and in 1939, 51 meetings were held at which 776 subjects were discussed (1976: 65, emphasis added).
2. This distinction is crucial for the recognition of landed property in the context of the 1970s oil nationalizations.
3. The original basing point, established at the Gulf of Mexico, was centered on the wellhead cost of US oil. The phantom freight was the bizarre calculation of the cost of delivery of oil from the Gulf of Mexico to any destination in the world regardless of its location of production and its point of origin.
4. The Organization of Petroleum Exporting Countries (OPEC) was formed in 1960. The original founders were Iran, Iraq, Kuwait, Saudi Arabia, and Venezuela.
5. Elwell-Sutton, L. P. (1955), *Persian Oil: The Study of Power Politics* (London: Lawrence and Wishart Publishing).
6. This is the same international energy agency whose emergence was anticipated and thus described in the US-UK Memorandum of 1964. Ironically, this "accuracy" also reveals the fossilized quality of US foreign policy in tandem with the As Is Achnacarry Agreement during the Pax Americana.
7. We believe that the era of Pax Americana ended in the late 1970s and with it the hegemony of its hegemon, the United States; see Bina 1993, 1994a, 1994c, 1994d, 1995, 1997. The end of Pax Americana means that hardly any of its remaining organizational arrangements or any of its lingering political paraphernalia can stand a chance for political (or ideological) legitimacy in the post-Pax Americana world. This includes the World Bank, NATO, and the IMF, to name a few.
8. Some writers saw this as a temporary setback for the United States. See, for instance, Bromley (1991; 205–8) and my review of it in Bina (1994c). On the US hegemony, besides the misconception of monopoly, a methodological trouble with Bromley is that it starts with the

conjunction of categories that are themselves in need of theoretical grounding.

9. The neoclassical theory of exhaustible resources contends that scarcity rent must be added to the marginal extraction cost of oil. This reflects the intertemporal opportunity cost of exploitation of oil, shown in terms of the rate of interest in Hotelling (1931). No concept of landed property can be found here, because "scarcity rent" is the measure of opportunity cost of intertemporal allocation. Hence, MC + Scarcity Rent = User Cost. I am hammering on this "formula" again and again to show the triviality of mainstream economic theory in its manifold configurations in the literature.
10. Some neoclassical economists assumed that all oil should be considered as produced at the time of discovery; hence oil reserves should be treated as inventory. Thus there is no room for rent other than market power and monopoly; see Adelman 1986, 1990.
11. For a critical review of competition see Bina 1985 (Ch. 6), 1989a; Clifton 1977; Semmler 1984; Shaikh 1980, 1982; Weeks 1981a (Ch. 6).
12. There is no shortage of vulgar interpretations of Marx's theory of rent in the oil literature. A recent invention can be found in Mommer 2002: 1–29, where he renames Marx's AR "customary ground rent" for payment of royalties. The rest of the volume is replete with ad hoc governance structures without any theoretical grounding with respect to the competitive unification of the global oil industry.
13. In the mid-1980s crisis, Texans told the following joke: "Do you know why Mercedes has no seat and no wheel this year?" The answer was: "Because the oilmen lost their behind and they don't know which way to turn." During this time the brief ploy of "swing production" by Saudi Arabia turned out to be a self-defeating project. On the one hand, withholding the production cut the total rent revenues via quantity. On the other hand, flooding the market (which is only possible for a limited time) cut the total rent revenues via price.
14. The sluggish decline of the oil prices, in the presence of excess supply, is not due to the orthodox presumption of market power; it is rather the consequence of the peculiarity of the oil production itself.
15. As has argued throughout this book, even slight mischaracterization of the post-1973–74 oil crisis (such as the one by Fine and Harris 1985) can be fatal in recognition and thus understanging of the competitive globalization of oil since the 1970s. The early 1970s oil crisis led, inter alia, to fanciful conspiracy views that gradually subsided by the durable reality of the globalization of oil. Yet, the residue of this uncritical, and indeed silly, approach to oil has not disappeared from the distorted imagination of the conspiracy buffs. The case in point is the recent revival by Nitzan and Bichler. They state: "Our analysis centers around *[sic] the process of differential capital accumulation*, emphasizing the quest to exceed the 'normal rate of return' and to expand one's share in the overall flow of profit" (1995: 446, emphasis in original). The authors mockingly

characterize the post-1973 oil crises as "energy conflicts" and emphasize that these prearranged and conspiratorial conflicts are the consequence of a "quest [by the oil companies] to exceed the 'normal rate of return' and to expand one's share in the overall flow of profit" (ibid: 446). They point out that their "methodological starting point" is the differential rate of return, albeit misconstrued as differential capital accumulation (Bichler and Nitzan 1996: 609), yet their result is also a differential rate of return. Thus, given the authors' repeated rendition of the conspiratorial (neoclassical) oil monopoly here and throughout their later works, neither capital accumulation nor state find any place in their tautological fantasyland. Yet, these two individuals have no hesitation to churn piece after piece, and to clobber dead horse after dead horse, with much impunity in the uncritical ambience and hype media frenzy in which an appeal to snake oil exceeds the penchant for genuine analysis. Likewise, it is also sadly astonishing to see that certain self-proclaimed Marxist authors (radical geographers) have been relying on parallel orthodox schemes, albeit in what appears as "post-modern" reasoning, claiming a brand-new radical explanation for oil, and the purported cause of the 2003 US invasion of Iraq (see Boal et al. 2005a, 2005b).

## 5 Oil and Capital: "Logic" of History and "Logic" of Territory

1. Karl Marx (1970), *A Contribution to the Critique of Political Economy* (Moscow: Progress Publishers), 210–13.
2. Vladimir Ilyich Lenin (1916), *Imperialism, the Highest Stage of Capitalism: A Popular Outline*: www.marxists.org/archive/lenin/works/1916/imp-hsc/.
3. See Cyrus Bina and Behzad Yaghmaian (1988), "Import Substitution and Export Promotion within the Context of the Internationalization of Capital," *Review of Radical Political Economics* 20 (2 & 3), Summer: 234–41; Cyrus Bina and Behzad Yaghmaian (1991), "Postwar Global Accumulation and the Transnationalization of Capital," *Capital & Class* 43, Spring: 107–30.
4. There is a vast literature on imperialism, among which Willoughby (1986) and Callinicos (2009) are interesting variations.
5. Nikolai Ivanovich Bukharin (1929), *Imperialism and World Economy* (New York: Monthly Review Press).
6. It should be duly noted that the historical context in which Lenin writes is of paramount consequence. The motivation for Lenin was primarily to combat the reformist position of Karl Kautsky—the then leader of the Second International—and the like. See V. I. Lenin (1970), *The Proletarian Revolution and the Renegade Kautsky* (Peking: Foreign Language Press). Therefore, to this extent, we wholeheartedly support Lenin's political position. Yet, we think that the by-product of this

rather lofty project turns out to be incompatible with Marx's methodology. Marx's theory of competition and value formation, is a grand project that pertains to the entire capitalist mode of production, regardless of its phases of development. Apart from a specific period of monopolies and cartels, the cardinal sin of relying on this pamphlet is to deny the application of the law of value for developed capitalism, and replacing it with a bourgeois concept of monopoly. After all, it would be silly for Marx to have gone through such painstaking methodological challenges and breathtaking theoretical questions if and only if the dynamics of capitalism, particularly in its more developed form of today, could be simply explicated by focusing on a few capitalists who tend to cartelize everything. Let us, once and for all, leave this emasculated and untrue worldview to bourgeois economists whose notion of competition is a (fictional) departure from the reality of concentration and centralization of capital. For the origin of Marxian theory of competition, see Karl Marx (1973b), *The Poverty of Philosophy* (Moscow: Progress Publishers), 126–34.

7. For periodization of the oil industry, analysis of the oil cartel, and a complete investigation of (Marxian) *value* and *competition* in this highly concentrated oil sector see Cyrus Bina (1985), *The Economics of the Oil Crisis* (New York: St. Martin's); for analysis of the global energy industry as a whole see Cyrus Bina (1989a), "Competitition, Control and Price Formation in the International Energy Industry," *Energy Economics* 11 (3), July.
8. See Cyrus Bina and Fernando Dachevsky (2008), "Bubbles, Risk, Crunch, and War," *Asia Times*, June 21.
9. The concept of fetishism is one of Marx's major contributions in *Capital*, vol. 1 (New York: Vintage Edition). Commodity fetishism refers to a process whereby the social relation among people turns into relation among their alienated labors in inanimate objects, namely, commodities. We contend that the erroneous and anachronistic perception of imperialism by the radical left should fall into the same category. They all take the ghostly appearance of socioeconomic/sociopolitical characteristics of the past epoch (i.e., epoch of imperialism) at its face value. Hence, the epiphenomenon of *imperial fetish* or, in the case of "post-modernists" *empire fetish*, in late capitalism. See Negri and Hardt (2000) and Hardt and Negri (2004) with respect to the latter case. These authors climbed to the height of *idealism*, sanguinely bordering on fairytale, against the very realities of our time.
10. Capitalism creates its own means as, for instance, the limitation of the "working day" has been overcome by the acceleration of technology and an increase in the productivity of labor through technological change.
11. Unfortunately, this very simple point has not yet been adequately understood by the liberal/radical left, for instance, on the issues of the US invasion of Iraq in which the left, in spite of existing evidence, alleges that the cause was the oil.

12. The case in point is the infamous "axis of evil" that was prepositioned by the Bush-Cheney administration for the invasion and occupation of Iraq, and kept rather ineptly as an excuse for the war with Iran. As we have shown elsewhere, neither the oil, "freedom," "democracy," "terrorism," nor the "defending a way of life," and so on, was the cause of the US invasion of Iraq. Indeed, as we persistently argued, this action proved to be detrimental to the global oil, global capital, and generally to the contemporary of globalization. To be sure, the 2003 US invasion of Iraq finds its origin in the neoconservative vision of wholesale demolition of the Middle East, a giant step for creating and securing a larger Israel. This was the real project and the real deal. Likewise, the 2006 Israeli invasion of Lebanon was the seeming botched-up drop of the second shoe with respect to the same project, so to speak. Oil was just the gravy that, according to original plan, would have comfortably bankrolled this undertaking. In consequence, not only liberals but also an assortment of leftists worldwide, including many self-styled Marxists, did voluntarily and/or involuntarily contribute to one of the most captivating campaigns of misinformation in the history of the United States. See Cyrus Bina (2004b), "The American Tragedy: The Quagmire of War, Rhetoric of Oil, and the Conundrum of Hegemony," *Journal of Iranian Research and Analysis* 20 (2), November; Cyrus Bina (1994b), "Oil, Japan and Globalization," *Challenge: The Magazine of Economic Affairs* 37 (3), May/June; Cyrus Bina (2006), "The Globalization of Oil—A Prelude to a Critical Political Economy," *International Journal of Political Economy* 35 (2), Summer.
13. The phrase "imperfect competition" was initially coined in Joan Robinson's book under the identical title in 1933. Virtually an identical version of this conception, "monopolistic competition," was independently developed by Edward H. Chamberlin as a PhD thesis at Harvard (defended in 1927) and published at the same time as Robinson's. See Joan Robinson (1969), *The Economics of Imperfect Competition* (London: Macmillan); Edward H. Chamberlin (1933) *Theory of Monopolistic Competition* (Cambridge, MA: Harvard University Press). It would be also instructive to revisit Joseph Schumpeter's (1954) monumental volume, *History of Economic Analysis* (New York: Oxford University Press), for a glimpse at the backdrop and ambience of the time in which these two scholars produced these books.
14. The fact that many orthodox as well as heterodox economists abandoned the "standard theory" and changed their focus on "theory of games" attests to the futility of this framework. Such a shift, however, has not been without problems of its own, not the least of which is the axiomatic behavior of interacting agents in isolation from the real structure. However, the principal indictment of the "theory of games" can be obtained from rent theory and its competitive formation within Marx's value theory. As we have demonstrated earlier, the magnitude of differential oil rents across the globe is not arbitrary but price-determined;

these rents are permanent, not transitory, and their magnitude is neither subject to a game-theoretic through bargaining by OPEC nor produced by monopoly pricing. These magnitudes are established prior to any bargaining mechanism that might be artificially at work. (see Bina 1985, Ch. 5, Ch.6; Bina 2012a; Bina 2013).

15. In the context of oil, the reader is alerted to the fact that Marx's theory competition in mature capitalism is also valid in the presence of rent. For application of competition, in the presence of a specific theory of oil rent, in the oil and energy industry see Bina 1985, Bina 2012a, Ch. 4.
16. The division of the world among monopolies, and the upsurge of finance capital in conjunction with it, in Lenin's *Imperialism* is illustrative of an early phase of capitalism in which nearly three-fourths of humanity had neither known what capitalism (i.e., a historical-specific social relation) means nor lived under it. In other words, in the epoch of imperialism, described rather specifically by Lenin, capitalism was not on its own. To be sure, while the circuits of capital in its commodity and finance forms were internationalized, the complete social circuit of capital (including the circuit of productive capital) has yet to be developed for, what Marx called, a complete "conquest of mode of production" via global competition. However, allusions, such as "decaying capitalism," must be attributed to the sentiments and anxieties at the time. Many revolutionaries at this juncture believed that the world revolution is at the door. For instance, one can be reminded of the Rosa Luxemburg's "theory of breakdown" in capitalism. Therefore, one should take this with a grain of salt and treat this period as a phase in development of world capitalism. By the same token, it is also imperative to suspect the reincarnation and currency of similar sentiments, for instance, as "new imperialism" (see Harvey 2003 and Harvey 2010). Harvey's *New Imperialism* is contrary to Marx's theory of competition; and, unlike Sweezy's, it is an oblique thesis on monopoly capital—a masterful smokescreen for further mystification of the present (see Bina 2013).
17. David Harvey (2011), "The Urban Roots of Financial Crises: Reclaiming the City for Anti-Capitalist Struggle," in Leo Panitch, Gregory Albo, and Vivek Chibber (eds.), *Socialist Register,* 1–35 (London: Merlin Press).
18. Harvey's ad hoc view of monopoly and monopolization does not allow for appreciation of the above average "organic composition of capital" in the oil sector, which is the effect of the mobility of capital associated with the intersectoral competition of capital beyond the geographical limits. He stubbornly clings to a duality of "territory" (i.e., geography) and "capital" (i.e., social relation), particularly when it comes to oil and oil fields, and then runs with it a circle of tautology to prove that the alleged OPEC monopoly and recent US interventions are connected and the world has now entered into an age of "The New Imperialism" (Harvey, 2003). See Fine (2006: 142) and Bina (2012a, Prologue and

p. 110) for a critique of this topsy-turvy observation that steeped in palpable misrepresentation and without much authenticity.

## 6 The Globalization of Energy

1. Competition within an industry brings about the formation of a market value along with unequal profit rates, whereas interindustry competition corresponds with the formation of different prices of production along with a tendency to develop a general rate of profit for the entire economy (Marx, *Capital* vol. 3, Ch. 10).
2. There is another category of rent, called absolute rent, which is dependent on the barrier of landed property. However, its actual manifestation in the energy industry is open to various interpretations.
3. Massarrat and I were the first, albeit independently of one another, to analyze the oil crisis of the early 1970s, and indeed theorize the oil and energy industry, in terms of the complex interaction of capital and the landed property via Marx. He has focused on US coal while I zeroed in on US oil to address the globalization of oil and energy industry through the epicenter of the oil crisis.
4. Another glitch in Massarrat's approach to rent is an arbitrary designation of price of oil by focusing on the market price of its final derivatives (in Nore's words: "final price of oil to consumers" [1980: 70]). Again, this arbitrary switching, which leads to the conflation of several different production processes, produces confusion as to what rent really is. In other words, once the crude oil is valorized, sold, and has left the market, it no longer should abide by the Marxian rent and the realm of landed property. Otherwise, we are back to the bourgeois notion of rent as market power. J. M. Chevalier (discussed in Bina 1989b: 95–97) is also trapped in this orthodox conundrum. Chevalier relies on monopoly rent and four different types of differential rents, namely, (a) quality rent, (b) position rent, (c) mining rent, and (d) technological rent; for a contrary view see Bina 1985, 1989b, 1992.

## 7 War, Oil, and Conundrum of Hegemony

1. For the preliminary assessment of US invasion of Iraq in 2003 see Bob Woodward (2004), *Plan of Attack* (New York: Simon & Schuster); Ron Suskind (2006), *The One Percent Doctrine* (New York: Simon & Schuster); James Risen (2006), *State of War* (New York: Free Press).
2. Kenneth Adelman (2002), "Cakewalk in Iraq," *The Washington Post*, February 13: A27. This is one of the first op-eds planted by an arch neocon for preparing the public for war on behalf of the Bush-Cheney administration. For a critic see Seymour Hersh (2004), *Chain of Command:*

*The Road from 9/11 to Abu Ghraib* (New York: HarperCollins). Adjunct to this unwarranted rush to war, there has been even a darker side—torture; see Jane Mayer (2008), *The Dark Side* (New York; Doubleday); Philippe Sands (2008), *Torture Team: Rumsfeld's Memo and the Betrayal of American Values* (New York: Palgrave/Macmillan).

3. See C. Keysen (1949), "Basing-Point Pricing and Public Policy," *Quarterly Journal of Economics* 62 (3): 289–314; G. Means (1972), "The Administered Price Thesis Confirmed," *American Economic Review* June; Fritz Machlup (1949), *The Basing-Point System* (Philadelphia, PA: Blackstone); John Blair (1976), *The Control of Oil* (New York: Pantheon); Cyrus Bina (1985-), *The Economics of the Oil Crisis* (London and New York: Merlin and St. Martin's).
4. See chapter 4 in this volume.
5. OPEC differential oil rents are being formed in competition. We rely on the Marxian and Schumpeterian competition, not the axiomatic fiat of competition that is often caricatured in neoclassical economics textbooks. Schumpeter, following Smith, Ricardo, and Marx, contends that competition is coercion of capital against capital, thus leading to further concentration and centralization and hence "strikes not at the margins of the profits and the outputs of the existing firms but at their foundations and their very lives." J. A. Schumpeter (1942), *Capitalism, Socialism and Democracy* (New York: Harper & Row), 84.
6. See chapter 4 in this volume.
7. See Cyrus Bina (1991), "War over Access to Cheap Oil or the Reassertion of U.S. Global Hegemony," in G. Bates (ed.), *Mobilizing Democracy: Changing the U.S. Role in the Middle East* (Monroe, Maine: Common Courage Press); Cyrus Bina (1993), "The Rhetoric of Oil and the Dilemma of War and American Hegemony," *Arab Studies Quarterly* 15 (3): 1–20.
8. To be sure, the US hope for domination of Iraq and establishment of a puppet government opens the door for the direct appropriation of Iraqi oil rents (revenues), not the determination of oil prices or "control" of oil.
9. The tragic events of September 11, 2001, are but the outward reflection of violent political and ideological forces that have long been accumulating in the American political outposts and thus ready for sudden internalization. Here, there is no distinction between "inside" and "outside" in the borderless world of globalization. See Immanuel Wallerstein (2002), "The Eagle Has Crash Landed," *Foreign Policy*, July/August: www.foreignpolicy.com/issue_julyaug_2002/ wallerstein.ht.
10. Cyrus Bina (1993), "The Rhetoric of Oil and the Dilemma of War and American Hegemony," *Arab Studies Quarterly* 15 (3): 1–20.
11. Michael T. Klare (2003), "It's the Oil, Stupid," *The Nation*, May 12; Gore Vidal (2002), "Blood for Oil," *The Nation*, October 28.
12. See Blair (1976), chapters 3, 4, and 5.

13. Cyrus Bina (1988). "Internationalization of the Oil Industry: Simple Oil Shocks or Structural Crisis?"
*Review: A Journal of the Fernand Braudel Centre* 11 (3): 329–79; Cyrus Bina (2006), "The Globalization of Oil: A Prelude to a Critical Political Economy," *International Journal of Political Economy* 35 (2): 4–34; Cyrus Bina (2013), "Synthetic Competition, Global Oil, and the Cult of Monopoly," in J. Moudud, C. Bina, and P. Mason (eds.), *Alternative Theories of Competition: Challenges to the Orthodoxy* (New York: Routledge).
14. See chapter 4 in this volume.
15. See James Schlesinger (1991), "Interview: Will War Yield Oil Security?" *Challenge,* March/ April. The right-wing view presented in this interview has since become a demarcation line for left-wing liberals when they speak of oil and war.
16. See Cyrus Bina (2008b), "Oil, War, Lies and 'Bullshit'," *Asia Times*, October 9, for a critique of need for "access" and direct control of oil reserves as a pretext for war; Cyrus Bina (2008a), "'Cakewalk' of Shame and Wickedness: *Mis*reading of History and the End of Lolita in Baghdad," *Negah*, June (in Persian): http://www.negah1.com/negah/negah22/negah22–21.pdf.
17. Michael T. Klare (2002), "Oiling the Wheels of War," *The Nation*, October 7. Klare's position on oil is an exact copy of Schlesinger's.
18. Michael T. Klare (2001), *Resource Wars* (New York: Henry Holt & Company), 15–16.
19. Michael T. Klare (2004), *Blood and Oil: The Dangers and Consequences of America's Growing Dependency on Imported Petroleum* (New York: Metropolitan Books), xiii.
20. See Cyrus Bina (2007), "America's Bleeding 'Cakewalk'," *EPS Quarterly* 19 (1) for a critique of US foreign policy and its connection with the Bush-Cheney administration political and ideological base in three elements, namely, neoconservatives, Cold Warriors, and Christian Zionist fundamentalists.
21. Edward N. Krapels (1993), "The Commanding Heights: International Oil in a Changed World," *International Affairs* 69 (1), January, is a right-wing piece with a penchant for returning to colonial era. In the present period, the wormhole of history is not only frequented by antiquated ideology (that is, religious fundamentalism and the like) but also by the residue of our immediate past that cries for civility and at the same time craves for annexation, unabashed and in "modern" secular vernacular.
22. Simon Bromley (1991), *American Hegemony and World Oil* (University Park, PA: The Pennsylvania State University), and author's review (1994c) in *Harvard Middle Eastern and Islamic Review* 1 (2): 194–198.
23. Klare 2003: 54. Klare's view of oil (and resources in general) is a remarkable illustration of commodity fetishism and thus distortion of social relations of capitalism.

24. According to Nordhaus estimates, the direct and indirect costs of forceful occupation of Iraq would range somewhere between $120 billion and $1.6 trillion over a 10-year period. William D. Nordhaus (2002), "Iraq: The Economic Consequences of War," *New York Review of Books* 49 (19), December 5; see Joseph E. Stiglitz and Linda Bilmes (2008), *The Three Trillion Dollar War: The True Cost of the Iraq Conflict* (New York: W. W. Norton).
25. Klare 2003.
26. See Paul Krugman (2003), "Nothing for Money," March 14: http://www.wwsprinceton.edu/~pkrugman/oildollar.html.
27. Antonio Gramsci (1971), *Selected from the Prison Notebooks* (New York: International Publishers).
28. Contrary the orthodoxy speaking of hegemony in fractured entities, such as "military hegemony," "political hegemony," or "economic hegemony," betrays the organic meaning of the concept. Again, many on the left are not in the listening mode.
29. For the Iran-Iraq war, see US Congress, Senate Committee on Foreign Relations (1987), *War in the Persian Gulf: The U.S. Takes Sides* (Washington, DC: US Government Printing Office), November; US Congress, House Committee on Foreign Relations (1990), Hearing before the Subcommittee on Europe and the Middle East, *United States-Iraq Relations* (Washington, DC: US Government Printing Office), April 26; The National Security Archive (2003), "Shaking Hands with Saddam Hussein: The U.S. Tilts Toward Iraq, 1980–1984," *National Security Archive Electronic Briefing Book* No. 82, Joyce Battle (ed.), February 25: http://www.gwu.edu/~nsarchiv/NSAEBB/NSAEBB82/ index.htm. The failure of Saddam Hussein (and, by proxy, the US government), to accomplish his mission of nipping Iran's militant Islamic government in the bud, led to a somewhat broader strategy that was first "theorized" by Samuel Huntington (1993) in "The Clash of Civilizations," *Foreign Affairs*, Summer. This, in our opinion, was the preamble to the Wolfowitz-Perle project that attempted rather successfully to create a "permanent war," in the Middle East.
30. See Perry Anderson (2002), "Force and Consent," (Editorial), *New Left Review* 17, September/October.
31. Cyrus Bina (1993), "The Rhetoric of Oil and the Dilemma of War and American Hegemony," *Arab Studies Quarterly* 15 (3): 1–20; Cyrus Bina (2004a), "Is It the Oil, Stupid?" *URPE Newsletter* 35 (3): 5–8; Cyrus Bina (2004b), "The American Tragedy: The Quagmire of War, Rhetoric of Oil, and the Conundrum of Hegemony," *Journal of Iranian Research and Analysis* 20 (2): 7–22.
32. See Ronald Steel (1967), *Pax Americana* (New York: Viking Press).
33. For a quick look, see Theodore Draper (1991), "American Hubris," in M. L. Sifry and C. Cerf (eds.), *The Gulf War Reader* (New York: Random House), 40–56.

34. See Melvyn P. Leffler (1984), "The American Conception of National Security and the Beginning of the Cold War, 1945–1948," *American Historical Review* 89: 346–81; D. F. Fleming (1961), *The Cold War and Its Origin*, 2 vols. (New York: Doubleday).
35. The 1953 and 1954 coup d'états against Mossadegh in Iran and Arbenz in Guatemala are the two prime examples of American subversion of democracy during the ascending years of Pax Americana. See Stephen Kinzer (2003), *All the Shah's Men: An American Coup and the Roots of Middle East Terror* (Hoboken, NJ: John Wiley & Sons); David Green (1971), *The Containment of Latin America* (Chicago, IL: Quadrangle Books).
36. See Cyrus Bina and Behzad Yaghmaian (1988), "Import Substitution and Export promotion Within the Context of the Internationalization of Capital," *Review of Radical Political Economics* 20 (2 & 3).
37. See Murray B. Levin (1971), *Political Hysteria in America: The Democratic Capacity for Repression* (New York: Basic books).
38. Many of these land reform programs were neither successful nor adequate for the construction of new social formation. Yet, they proved mightily successful for their irreversible destruction of the old. See Cyrus Bina and Behzad Yaghmaian (1991), "Postwar Global Accumulation and the Transnationalization of Capital," *Capital & Class* 43.
39. See Robert Bocock (1986), *Hegemony* (London: Tavistock Publications). This is an excellent source for the analysis of Gramsci's notion of hegemony and its multifaceted applications to politics and social analysis. It also provides a hint of criticism of Althusser on "political society" in contradistinction with "civil society." See also Perry Anderson, "The Antinomies of Antonio Gramsci," *New Left Review* 100 (1976–77); A. Showstack Sassoon (1982), "Passive Revolution and the Politics of Reform," in Sassoon (ed.), *Approaches to Gramsci* (London: Writers and Readers); Peter D. Thomas (2009), *The Gramscian Moment* (Chicago, IL: Haymarket Books).
40. See Cyrus Bina (1994a), "Farewell to the *Pax Americana*: Iran, Political Islam, and the Passing of the Old Order," in H. Zangeneh (ed.), *Islam, Iran, and World Stability* (New York: St. Martin's Press), 41–74.
41. To be sure, the Wolfowitz-Perle project was in place long before September 11, 2001. See http:// www.cooperativeresearch.org/organization/profiles/ defensepolicyboard.html.
42. White House (2002), *The National Security Strategy of the United States of America* (Washington, DC: Government Printing Office), September 17.
43. National Security Decision Directive (NSDD 99), July 12, 1983, November 21, 1983 memo ("Iraqi Use of Chemical Weapons") sent by Jonathan T. Howe to Lawrence S. Eagleburger; December 10, 1983, cable, concerning "Rumsfeld Visit to Iraq," to the State Department from US Interest Section in Iraq; December 21, 1983, cable from US Embassy in UK to the State Department concerning Rumsfeld

meeting with Saddam Hussein. See Joyce Battle (ed.), "Shaking Hands with Saddam Hussein: The U.S. Tilts Toward Iraq, 1980–1984," The *National Security Archive*, National Security Archive Electronic Briefing Book No. 82, February 25, 2003: http://www.gwu.edu/~nsarchiv/ NSAEBB/ NSAEBB82/index.htm.

44. *The National Security Archive*. p. 3, emphasis added.
45. *The National Security Archive*. pp. 3–4, emphasis added. For more on Iran see Bina and Zangeneh (1992), Zangeneh (1994), Bina (1999), Bina (2009a), Bina (2009b), Bina (2010a), Bina (2010b), Peterson (2012a), Peterson (2012b), Schell (2012), Gladstone and Sanger (2012), Erdbrink (2012), Leverett and Leverett (2012), Bauer (2012), Crist (2012), Hersh (2012).
46. Hans Blix (2004), *Disarming Iraq* (New York: Pantheon Books).
47. The allegation of WMD is turned out to be a ploy, given that the Bush administration was not interested in UN inspections all along. See Blix 2004.
48. According to BBC, Powell, following his presentation, indicated: "I would call my colleagues' attention to the fine paper that United Kingdom distributed yesterday, which describes in exquisite detail Iraqi deception activities." The British report copied (including the typos) from Ibrahim Al-Marashi (2002), "Iraq's Security and Intelligence Network: A Guide and Analysis," *Middle East Review of International Affairs* 6 (3), September , which has nothing to do with the 2003 allegation of WMD in Iraq.
49. See *Secretary of State Addresses the UN Security Council*, February 5, 2003: http://www.whitehouse.gov/news/ releases/2003/02/20030205-1.html.
50. Douglas Jehl (2004), "U.S. Report Finds Iraqis Eliminate Illicit Arms in 90's," *New York Times*, October 7: 1, 22.
51. See Richard Clarke (2004), *Against All Enemies* (New York: Free Press).
52. See Chris Hedges (2006), *American Fascists: The Christian Right and the War on America* (New York: Free Press); Michelle Goldberg (2007), *Kingdom Coming: The Rise of Christian Nationalism* (New York: W. W. Norton & Company); Michael Lackey (2012), *The Modernist God State—A Literary Study of the Nazis Christian Right* (New York: Continuum International Publishing Group).

# Bibliography

Aaronson, Trevor (2011). "Terrorists for the FBI." *Mother Jones* September/ October: 30–43.

Adelman, Kenneth (2002). "Cakewalk in Iraq." *The Washington Post* February 13: A27.

Adelman, M. A. (1992) "Finding and Development Costs in the United States, 1945–1986." In J. R. Moroney (ed.), *Advances in the Economics of Energy and Resources*, vol. 7. Greenwich, CT: JAI Press.

———. (1972). *The World Petroleum Market*. Baltimore, MD: Johns Hopkins University Press.

———. (1980). "The Clumsy Cartel." *Energy Journal* 4 (3).

———. (1982). "OPEC as a Cartel." In J. M. Griffin and D. J. Teece (eds.), *OPEC Behavior and World Oil Prices*. London: Allen and Unwin.

———. (1986). "Scarcity and World Oil Prices." *Review of Economics and Statistics* 68: 387–97.

———. (1990). "Mineral Depletion, with Special Reference to Petroleum." *Review of Economics and Statistics* 72: 1–10.

———. (1991) "Oil Fallacies." *Foreign Policy* 82, Spring: 2–16.

Adelman, M. A. and M. Shahi (1989). "Oil Development-Operating Cost Estimates, 1955–1985." *Energy Economics* 11: 2–10.

Akins, James E. (1973). "The Oil Crisis: This Time the Wolf is Here." *Foreign Affairs* 51 (April): 462–90.

Alhajji, A. F. and D. Huettner (2000). "OPEC and World Crude Oil Markets from 1973 to 1994: Cartel, Oligopoly, or Competitive?" *The Energy Journal* 21 (3): 31–60.

Alfonso, P. J. P. (1966). "The Organization of Petroleum Exporting Countries." *Monthly Bulletin* (Ministry of Mines and Hydrocarbons, Caracas) 1 (1–4).

Al-Marashi, Ibrahim (2002). "Iraq's Security and Intelligence Network: A Guide to Analysis." *Middle East Review of International Affairs* 6 (3), September.

Alnasrawi, A. (1973). "Collective Bargaining Power in OPEC." *Journal of World Trade Law* 7 (2).

———. (1985). *OPEC in a Changing World Economy*. Baltimore, MD: John Hopkins University Press.

American Petroleum Institute (1979). *Basic Petroleum Data Book*. Washington, DC: API.

Amin, Samir (1974). *Accumulation on a World Scale.* New York: Monthly Review Press.

Anderson, J. (1797). *Essays Relating to Agriculture and Rural Affairs.* Dublin: P. Wogan, P. Byrne, and J. Moore.

———. (1777). *Observations on the Means of Exciting a Spirit of National Industry,* Reprints of Economic Classics, 1968. New York: Augustus M. Kelley Publishers.

Anderson, Jack and Les Whitten (1977). "Kissinger Allowed Oil Prices to Soar." *The Washington Post* July 7, D15.

Anderson, Perry (1976–77). "The Antinomies of Antonio Gramsci." *New Left Review* 100.

———. (2002). "Force and Consent." (Editorial), *New Left Review* 17, September/October.

Aperjis, D. (1982). *The Oil Market in the 1980s: OPEC Oil Policy and Economic Development.* Cambridge, MA: Ballinger.

Arrighi, Giovanni (1994). *The Long Twentieth Century.* London: Verso.

Ball, G. W. (1965). "Circular Aerogram 5671: Energy Diplomacy and Global Issues." In *Foreign Relations of the United States 1964–1968* Vol. 34 (1999): 333. Washington, DC: Government Printing Office.

Baran, Paul and Paul Sweezy (1966). *Monopoly Capital.* New York: Monthly Review Press.

Barrow, J. D. (1991). *Theories of Everything.* New York: Oxford University Press.

Battle, Joyce (2003). "Shaking Hands with Saddam Hussein: The U.S. Tilts Toward Iraq, 1980–1984." *National Security Archive Electronic Briefing Book* 82, February 25: www.gwu.edu/~nsarchiv/NSAEBB/NSAEBB82/index.htm.

Bauer, Shane (2012). "No Way Out." *Mother Jones* 37 (6), November/December: 22–32, 67.

Beinart, P. (2012). *The Crisis of Zionism.* New York: Times Books.

Benton, Ted (1984). *The Rise and Fall of Structural Marxism: Althusser and His Influence.* New York: St. Martin's.

Bichler, S. and J. Nitzan (1996). "Putting the State in Its Place: U.S. Foreign Policy and Differential Capital Accumulation in Middle East 'Energy Conflicts.'" *Review of International Political Economy* 3 (4): 608–61.

Bina, C. (1985). *The Economics of the Oil Crisis.* London and New York: The Merlin Press and St. Martin's Press.

———. (1988). "Internationalization of the Oil Industry: Simple Oil Shocks or Structural Crisis?" *Review: A Journal of Fernand Braudel Center* 11 (3): 329–79.

———. (1989a). "Competition, Control and Price Formation in the International Energy Industry." *Energy Economics* 11 (3): 162–68.

———. (1989b). "Some Controversies in the Development of Rent Theory: The Nature of Oil Rent." *Capital and Class* 39: 82–112.

———. (1990). "Limits to OPEC Pricing: OPEC Profits and the Nature of Global Oil Accumulation." *OPEC Review* 14 (1): 5–73.

———. (1991). "War over Access to Cheap Oil or the Reassertion of U.S. Global Hegemony." In G. Bates (ed.), *Mobilizing Democracy: Changing the U.S. Role in the Middle East.* Monroe, ME: Common Courage Press.

———. (1992). "The Law of Economic Rent and Property." *American Journal of Economics and Sociology* 51 (2): 187–203.

———. (1993). "The Rhetoric of Oil and the Dilemma of War and American Hegemony." *Arab Studies Quarterly* 15 (3): 1–20.

———. (1994a). "Farewell to the *Pax Americana*: Iran, Political Islam, and the Passing of the Old Order." In H. Zangeneh (ed.), *Islam, Iran, and World Stability,* 41–74. New York: St. Martin's Press.

———. (1994b). "Oil, Japan and Globalization." *Challenge: The Magazine of Economic Affairs* 37 (3), May/June.

———. (1994c). "Review of American Hegemony and World Oil." *Harvard Middle Eastern and Islamic Review* 1 (2): 194–98.

———. (1994d). "Towards a New World Order." In H. Mutalib and T. Hashmi (eds.), *Islam, Muslims and the Modern State*, 3–30. London: Macmillan.

———. (1995). "On Sand Castles and Sand-Castle Conjectures: A Rejoinder." *Arab Studies Quarterly* 17 (1–2): 167–71.

———. (1997). "Globalization: The Epochal Imperatives and Developmental Tendencies." In D. Gupta (ed.), *The Political Economy of Globalization*, 41–58. Boston, MA: Kluwer.

———. (1999). "Hot Summer of Defiance: The Student Protests for Freedom and Democracy in Iran." *Journal of Iranian Research and Analysis* 15 (2): 47–60.

———. (2004a). "Is It the Oil, Stupid?" *URPE Newsletter* 35 (3): 5–8.

———. (2004b). "The American Tragedy: The Quagmire of War, Rhetoric of Oil, and the Conundrum of Hegemony." *Journal of Iranian Research and Analysis* 20 (2): 7–22.

———. (2006). "The Globalization of Oil: A Prelude to a Critical Political Economy." *International Journal of Political Economy* 35 (2), Summer.

———. (2007). "America's Bleeding 'Cakewalk'." *EPS Quarterly* 19 (1) March.

———. (2008a). "'Cakewalk' of Shame and Wickedness: *Mis*reading of History and the End of Lolita in Baghdad." *Negah* June (in Persian): http://www.negah1.com/negah/negah22/negah22-21.pdf.

———. (2008b). "Oil, War, Lies and 'Bullshit'." *Asia Times*, October 9.

———. (2009a). "Petroleum and Energy Policy in Iran." *Economic and Political Weekly* 44 (1).

———. (2009b). "Iran: Crossroads of History and a Critique of Prevailing Political Perspectives." *Journal of Iranian Research and Analysis* (Special Issue on Post-Election Iran), 26 (2): Fall.

———. (2010a). "US Foreign Policy and Post-Election Iran: An Open Letter to President Obama." *CounterPunch* March 12.

———. (2010b). "Post-Election Iran and US Foreign Policy." *EPS Quarterly* 22 (1): March.

Bina, C. (2011). "Globalization, Value Theory and Crisis." In I. Siriner, et al. (eds.) *Political Economy, Crisis & Development*, 19–24. London: IJOPEC Publication.

———. (2012a). *Oil: A Time Machine: Journey Beyond Fanciful Economics and Frightful Politics.* Second Edition. Ronkonkoma, NY: Linus.

———. (2012b). "Who's Afraid of Kirchner's Oil Nationalization." *Asia Times* May 8.

———. (2012c). "Oil." In M. Juergensmeyer and H. K. Anheier (eds.), *Encyclopedia of Global Studies.* Thousand Oaks, CA: Sage Publishers.

———. (2013). "Synthetic Competition, Global Oil, and the Cult of Monopoly," In J. Moudud, C. Bina, and P. Mason (eds.), *Alternative Theories of Competition: Challenges to the Orthodoxy,* 55–85. New York and London: Routledge.

Bina, C. and Behzad Yaghmaian (1988). "Import Substitution and Export Promotion within the Context of the Internationalization of Capital." *Review of Radical Political Economics* 20 (2 & 3), Summer: 234–41.

———. (1991). "Postwar Global Accumulation and the Transnationalization of Capital." *Capital & Class* 43, Spring: 107–30.

Bina, C. and Chuck Davis (1996). "Wage Labor and Global Capital: Global Competition and Universalization of the Labor Movement." In C. Bina, L. Clements, and C. Davis (eds.), *Beyond Survival: Wage Labor in the Late Twentieth Century.* Armonk, NY: M. E. Sharpe.

———. (2008). "Contingent Labor and Omnipotent Capital: The Open Secret of Political Economy." *Political Economy Quarterly* 4 (15): 166–211.

Bina, C. and F. Dachevsky (2008). "Bubbles, Risk, Crunch, and War." *Asia Times* June 21.

Bina, C. and M. Vo (2007). "OPEC in the Epoch of Globalization: An Event Study of Global Oil Prices." *Global Economy Journal* 7 (1): 1–49.

Bina, C., L. Clements, and C. Davis, eds. (1996). *Beyond Survival: Wage Labor in the Late Twentieth Century.* Armonk, NY: M. E. Sharpe.

Bina, C. and H. Zangeneh, eds. (1992). *Modern Capitalism and Islamic Ideology in Iran.* New York: St. Martin's.

Blake, K. (2009). *The US-Soviet Confrontation in Iran, 1945–1962.* Lanham, MD: University Press of America.

Bleany, Michael (1976). *Underconsumption Theories: A History and Critical Analysis.* New York: International Publishers.

Blitzer, C., A. Meeraus, and A. Stoutjesdijk (1975). "A Dynamic Model of OPEC Trade and Production," *Journal of Development Economics* 2(4): 319–335.

Blair, J. M. (1976). *The Control of Oil.* New York: Pantheon Books.

Blix, Hans (2004). *Disarming Iraq.* New York: Pantheon Books.

Boal, I., T. J. Clark, J. Matthews, and M. Watts (2005a). *Afflicted Powers.* London: Verso.

———. (2005b). "Blood for Oil." *London Review of Books,* April 21.

Bobrow, D. and R. Kurdle (1975). "Theory, Policy and Resource Cartels: The Case of OPEC." *Journal of Conflict Resolution* 20 (1).

Bocock, Robert (1986). *Hegemony*. London: Tavistock Publications.
Brenner, Lenni (1984). *The Iron Wall: Zionist Revisionism from Jabotinsky to Shamir*. London: Zed Books.
Brenner, Robert (1977). "The Origins of Capitalist Development: A Critique of Neo-Smithian Marxism." *New Left Review* 104 (July–August): 25–92.
Brewer, Anthony (1980). *Marxist Theories of Imperialism: A Critical Survey*. London: Routledge & Kegan Paul.
BP Statistical Review (2012). *Energy Outlook 2030*. London, January.
Bromley, S. (1991). *American Hegemony and World Oil*. University Park, PA: Pennsylvania State University Press.
———. (2005). "The United States and the Control of World Oil." *Government and Opposition* 40 (2), Spring.
Brown, G. (1974). "An Optimal Program for Managing Common Property Resources With Congestion Externalities." *Journal of Political Economy* 82 (1).
Brown, H. G. (1941). "Economic Rent: In What Sense a Surplus." *American Economic Review* 31.
Bryan, Dick (1990). " 'Natural' and 'Improved' Land in Marx's Theory of Rent." *Land Economics* 66 (2), May: 176–81.
Bukharin, Nikolai Ivanovich (1929). *Imperialism and World Economy*. New York: Monthly Review Press.
Bush, George H. W. (1991). "State of the Union Message to the Nation," *New York Times* January 30.
Callinicos, Alex (2009). *Imperialism and Global Political Economy*. London: Polity.
Cattan, H. (1967a). *The Evolution of Oil Concessions in the Middle East and North Africa*. Dobbs Ferry, NY: Oceana Publications.
———. (1967b). *The Law of Oil Concessions in the Middle East and North Africa*. Dobbs Ferry, NY: Oceana Publications.
Chamberlin, Edward H. (1933). *Theory of Monopolistic Competition*. Cambridge, MA: Harvard University Press.
Chapman, D. (1983). *Energy Resources and Energy Corporations*. Ithaca, NY: Cornell University.
Chevalier, J. M. (1976). "Theoretical Elements for an Introduction to Petroleum Economics." In A. P. Jauemin and H. W. Dejong (eds.), *Market, Corporate Behavior, and the State*. The Hague: Martinus Nijhoff.
Clark, J. B. (1891). "Distribution as Determined by a Law of Rent." *Quarterly Journal of Economics* 5.
Clark, J. M. (1938). "Basing-point Methods of Price Quoting." *Canadian Journal of Economics and Political Science* 4.
Clifton, J. (1977). "Competition and Evolution of the Capitalist Mode of Production." *Cambridge Journal of Economics* 1 (2): 137–51.
Cremer, J. and D. Salehi-Isfahani (1980). "A Theory of Competitive Pricing in the Oil Market: What Does OPEC Really Do?" *CARESS*, University of Pennsylvania, Philadelphia, PA, Working Paper No. 80–4.

Cremer, J. and D. Salehi-Isfahani (1989). "The Rise and Fall of Oil Prices: A Competitive View" *Annales D'Economie etDe Statistique* 15 (16): 427–54.

Crist, D. (2012). *The Twilight War: The Secret History of America's Thirty-Year Conflict with Iran*. New York: Penguin.

Dasgupta, P. S. and G. M. Heal (1979). *Economic Theory and Exhaustible Resources*. Cambridge: Cambridge University Press.

Davenport, E. R. and S. R. Cooke (1923). *The Oil Trusts and Anglo-American Relations*. London: Macmillan.

Davidson, Paul (1979). "The Economics of Natural Resources." *Challenge* 22 (1), March/April.

De Chazeau, M. G. and A. E. Kahn (1959). *Integration and Competition in the Petroleum Industry*. New Haven, CT: Yale University Press.

DeNovo, J. (1956). "The Movement for an Aggressive American Oil Policy, 1918–1920." *American Historical Review* July.

Drezner, D. W. (2011). *Theories of International Politics and Zombies*. Princeton, NJ: Princeton University Press.

Devarajan, S. and A. C. Fisher (1982). "Exploration and Scarcity." *Journal of Political Economy* 90 (6): 1279–90.

Draper, Theodore (1991). "American Hubris." In M. L. Sifry and C. Cerf (eds.), *The Gulf War Reader*. New York: Random House.

Eichenwald, K. (2012). *500 Days: Secrets and Lies in the Terror Wars*. New York: Touchstone Books.

Elm, M. (1992). *Oil, Power, and Principle: Iran's Oil Nationalization and Its Aftermath*. Syracuse, NY: Syracuse University Press.

Elwell-Sutton, L. P. (1955). *Persian Oil: A Study of Power Politics*. London: Lawrence and Wishart.

Emmanuel, Arghiri (1972). *Unequal Exchange*. New York: Monthly Review Press.

Energy Information Administration. (1982a.) *Annual Report to Congress*, vol. 2. Washington, DC: US Government Printing Office.

———. (1982b). *Outlook for US Coal*. Washington, DC: US Government Printing Office.

Energy Information Administration (2010). *Annual Energy Review*. Washington, DC: US Government Printing Office.

Energy Information Administration (2012). *Annual Energy Review*. Washington, DC: US Government Printing Office.

Engler, R. (1961). *The Politics of Oil*. Chicago, IL: University of Chicago Press.

———. (1977). *The Brotherhood of Oil*. Chicago, IL: University of Chicago Press.

Erbrink, T. (2012). "As Iran Currency Keeps Tumbling, Anxiety is Rising." *New York Times* October 5: A1, A6.

Ezzati, A. (1976). "Future OPEC Price and Production Strategies as Affected by Its Capacity to Absorb Oil Revenue," *European Economic Review* 8: 107–38.

Federal Trade Commission (1952). *International Petroleum Cartel.* A Report to the Subcommittee on Monopoly, Select Committee on Small Business (82d Congress, 2d Session). Washington, DC: US Government Printing Office.

Fine, B. (1979). "On Marx's Theory of Agricultural Rent." *Economy and Society* 8 (3): 241–78.

———. (1982a) "Landed Property and the Distinction Between Royalty and Rent." *Land Economics* 58 (3).

———. (1982b). *Theories of the Capitalist Economy.* New York: Holmes & Meier Publisher.

———. (1983). "The Historical Approach to Rent and Price Theory Reconsidered." *Australian Economic Papers* 22 (40): 132–43.

———. (1986). "A Dissenting Note on the Transformation Problem." In B. Fine (ed.), *The Value Dimension: Marx versus Ricardo and Sraffa,* 209–14. London: Routledge & Kegan Paul.

———. (1990). *The Coal Question: Political Economy and Industrial Change from the Nineteenth Century to the Present Day.* London: Routledge.

———. (1994). "Coal, Diamonds, and Oil: Toward a Comparative Theory of Mining." *Review of Political Economy* 6 (3): 279–302.

———. (2006). "Debating the 'New' Imperialism." *Historical Materialism* 14 (4): 133–56.

Fine, B. and L. Harris (1979). *Rereading Capital.* London: Macmillan.

———. (1985). *The Peculiarities of the British Economy.* London: Lawrence & Wishart.

Fine, B. and A. Saad-Filho (2008). "Production versus Realization in Marx's Theory of Value: A Reply to Kincaid." *Historical Materialism* 16 (4): 191–204.

———. (2009). "Twixt Ricardo and Rubin: Debating Kincaid Once More." *Historical Materialism* 17, 3: 192–207.

———. (2010). *Marx's 'Capital'.* 5th ed. London: Pluto Press.

Fischer, D., D. Gately, and J. F. Kyle (1975). "The Prospects for OPEC: A Critical Survey of Models of the World Oil Market." *Journal of Development Economics* 2.

Fitch, B. (1982). "OPEC: The Big Cartel That Couldn't..." *Against the Current* 1 (4).

Fleming, D. F. (1961). *The Cold War and Its Origin,* vol. 2. New York: Doubleday.

Ford, A. W. (1954). *The Anglo-Iranian Oil Dispute of 1951–52.* Berkeley: University of California Press.

Foster, E. (1981). "The Treatment of Rents in Cost-Benefit Analysis." *American Economic Review* 71 (1).

Frank, Andre Gunder (1969a). *Capitalism and Underdevelopment in Latin America.* New York: Monthly Review Press.

———. (1969b). *Latin America: Underdevelopment or Revolution.* New York: Monthly Review Press.

Frank, Andre Gunder. (1972). *Lumpen Bourgeoisie, Lumpen Development*. New York: Monthly Review Press.
Frank, H. J. (1966). *Crude Oil Prices in the Middle East*. New York: Praeger Publishers.
Frankfurt, Harry G. (2005). *On Bullshit*. Princeton, NJ: Princeton University Press.
Frankel, P. H. (1961). *The Essentials of Petroleum*. London: Frank Cass.
Fukuyama, F. (1992). *The End of History and the Last Man*. New York: Avon Books.
Gasiorowski, M. (1987). "The 1953 Coup d'état in Iran." *International Journal of Middle East Studies* 19: 261–86.
Gasiorowski, M. and M. Byrne, eds. (2004). *Mohammad Mosaddeq and the 1953 Coup in Iran*. Syracuse, NY: Syracuse University Press.
Gately, D. (1984). "A Ten Year Retrospective: OPEC and the World Oil Market." *Journal of Economic Literature* 22 (3): 1100–14.
———. (1986). "The Prospects for Oil Prices, Revisited." *Annual Review of Energy* 11: 513–38.
George, H. (1938). *Progress and Poverty*. New York: Robert Schalkenbach Foundation.
Girvan, Norman (1975). "Economic Nationalism." *Daedalus* 104 (4): 145–58.
———. (1976). *Corporate Imperialism: Conflict and Expropriation*. New York: Monthly Review Press.
Gladstone, R. and D. Sanger (2012). "Nod to Obama by Netanyahu on Iran Bomb." *New York Times* September 28: A1, A12.
Gramsci, Antonio (1971). *Selected from the Prison Notebooks*. New York: International Publishers.
Gray, L. C. (1914). "Rent Under the Assumption of Exhaustibility." *Quarterly Journal of Economics* 28.
Green, David (1971). *The Containment of Latin America*. Chicago, IL: Quadrangle Books.
Greene, B. R. (2004). *The Fabric of the Cosmos*. New York: Vintage.
Greider, W. and J. P. Smith (1977). "A Proposition: High Oil Prices Benefit U.S." *The Washington Post* July 10, Al.
Griffin, J. M. (1985). "OPEC Behavior: A Test of Alternative Hypotheses." *American Economic Review* 75: 954–63.
Griffin, J. M. and D. J. Teece, eds. (1982). *OPEC Behavior and World Oil Prices*. London: Allen and Unwin.
Hanieh, A. (2011). "Finance, Oil and the Arab Uprisings: The Global Crisis and the Gulf States." In Leo Panitch, Gregory Albo, and Vivek Chibber (eds.), *Socialist Register 2012*, 176–99. London: Merlin Press.
Hardt, M. and A. Negri (2004). *Multitude: War and Democracy in the Age of Empire*. New York: Penguin.
Harcourt, G. C. (1972). *Some Cambridge Controversies in the Theory of Capital*. Cambridge :Cambridge University Press.
Harkinson, J. (2012). "Drill, Baby, Drill." *Mother Jones* 37 (6) November/December: 54–60, 66–67.

Hartshorn, J. E. (1967). *Oil Companies and Governments.* 2nd ed. London: Faber and Faber.

Harvey, D. (1999[1982]). *The Limits to Capital.* New York: Verso.

———. (2003). *The "New" Imperialism.* Oxford: Oxford University Press.

———. (2010). *The Enigma of Capital and the Crisis of Capitalism.* Oxford: Oxford University Press.

———. (2011). "The Urban Roots of Financial Crises: Reclaiming the City for Anti-Capitalist Struggle." In Leo Panitch, Gregory Albo, and Vivek Chibber (eds.), *Socialist Register 2012,* 1–35. London: Merlin Press.

Heal, G. and G. Chichilnisky (1991). *Oil and the International Economy.* Oxford: Clarendon Press.

Heinrich, Michael (2012). *An Introduction to the Three Volumes of Marx's Capital.* New York: Monthly Review Press.

Hersh, Seymour M. (2004). *Chain of Command: The Road from 9/11 to Abu Ghraib.* New York: HarperCollins

———. (2012). "Our Men in Iran." *New Yorker* April 6.

Hidy, R. W. and E. Muriel (1956). *Pioneering in Big Business: History of Standard Oil Co. (New Jersey) 1882–1911.* New York: Harper.

Hilferding, R. (1981). *Finance Capital: A Study of the Latest Phase of Capitalist Development.* London: Routledge & Kegan Paul.

Hirst, D. (1966). *Oil and Public Opinion in the Middle East.* New York: Praeger Publishers.

Hnyilicza, E. and R. Pindyck (1976). "Pricing Policies for a Two-Part Exhaustible Resource Cartel." *European Economic Review* 8: 139–54.

Hobson, J. A. (1891). "The Law of the Three Rents." *Quarterly Journal of Economics* 5: 263–88.

———. (1961 [1902]). *Imperialism—A Study.* London: George Allen & Unwin.

Hotelling, H. (1931). "The Economics of Exhaustible Resources." *Journal of Political Economy* 39 (2): 137–75.

Houthakker, H. S. (1976). *The World Price of Oil.* Washington, DC: American Enterprise Institute.

Hubbert, M. K. (1969). "Energy Resources." In National Academy of Sciences National Resource Council, *Resources for Man*, 157–242. San Francisco, CA: W. H. Freeman.

Huntington, S. (1993). "The Clash of Civilizations." *Foreign Affairs,* Summer.

Hyde, James N. (1956). "Permanent Sovereignty over Natural Wealth and Resources." *American Journal of International Law* 50 (4): 854–67.

Ise, J. (1926). *United States Oil Policy.* New Haven, CT: Yale University Press.

Issawi, C. P. and M. Yeganeh (1962). *Economics of Middle Eastern Oil.* New York: Praeger Publishers.

Jabotinsky, Vladimir (Ze'ev) (1937 [1923]). "The Iron wall." *The Jewish Herald* (South Africa) November 26.

Jacoby, N. H. (1974). *Multinational Oil: A Study of Industrial Dynamics.* New York: Macmillan.

Jehl, Douglas (2004). "U.S. Report Finds Iraqis Eliminate Illicit Arms in 90's." *New York Times* October 7: A1, A22.

Johany, A. D. (1980). *The Myth of the OPEC Cartel: The Role of Saudi Arabia*. New York: John Wiley.

Jones, C. (1990). "OPEC Behavior Under Falling Prices: Implications for Cartel Stability." *The Energy Journal* 11: 117–29.

Kalecki, M. (1954). *The Theory of Economic Dynamics*. New York: Rinehart.

Keynes, J. M. (1964). *The General Theory of Employment, Interest, and Money*. New York: Harcourt & Brace

Keysen, C. (1949). "Basing Point Pricing and Public Policy." *Quarterly Journal of Economics* 62 (3): 289–314.

Keysen, C. and D. F. Turner (1959). *Antitrust Policy—An Economic and Legal Analysis*. Cambridge, MA: Harvard University Press.

Kincaid, J. (2007). "Production versus Realization: A Critique of Fine and Saad-Filho on Value theory." *Historical Materialism* 15 (4): 137–65.

———. (2008). "Production versus Capital in Motion: A Reply to Fine and Sass-Filho." *Historical Materialism* 16 (4): 181–203.

Kinzer, Stephen (2003). *All the Shah's Men: An American Coup and the Roots of Middle East Terror*. Hoboken, NJ: John Wiley & Sons.

Kirzner, I. (1987). "Competition: Austrian Conceptions." In J. Eatwell, M. Milgate, and P. Newman (eds.), *The New Palgrave: A Dictionary of Economics*. London: Macmillan,.

Klare, M. (2001). *Resource Wars*. New York: Henry Holt & Company.

———. (2002). "Oiling the Wheels of War." *The Nation* October 7.

———. (2003). "It's the Oil, Stupid." *The Nation* May 12.

———. (2004). *Blood and Oil: The Dangers and Consequences of America's Growing Dependency on Imported Petroleum*. New York: Metropolitan Books.

Kolko, Gabriel (2002). *Another Century of War*. New York: The New press.

Kolm, Serge-Christophe (1981). "Liberal Transition to Socialism: Theory and Difficulties." *Review: A Journal of the Fernand Braudel Center* 5 (2): 205–18.

Kramer, Andrew E. (2012) *New York Times* June 14: http://www.nytimes.com/2012/06/15/business/global/opec-is-said-to-leave-production-steady.html.

Krapels, Edward N. (1993). "The Commanding Heights: International Oil in a Changed World." *International Affairs* 69 (1), January.

Krasner, S. D. (1978). *Defining the National Interest: Raw Materials Investments and U.S. Foreign Policy*. Princeton, NJ: Princeton University Press.

Krueger, A. (1974). "The Political Economy of the Rent-Seeking Society." *American Economic Review* 64 (3).

Krueger, R. B. (1975). *The United States and International Oil*. New York: Praeger Publishers.

Krugman, Paul (2003). "Nothing for Money." March 14: http://www.wwsprinceton.edu/~pkrugman/oildollar.html.

*Latin American Perspectives* (1976). "Capitalism in Latin America: Process of Underdevelopment." 3 (2), Issue 9, Spring.

———. (1977). "Peru: Bourgeois Revolution and Class Struggle." 4 (3), Issue 14, Summer.

———. (1979). "Views on Dependency." 6 (2), Issue 21, Spring.

———. (1981). "Dependency and Marxism." 8 (3–4), Issues 30–31, Summer and Fall.

Leeman, W. (1962). *The Price of Middle East Oil.* Ithaca, NY: Cornell University Press.

Leffler, Melvyn P. (1984). "The American Conception of National Security and the Beginning of the Cold War, 1945–1948." *American Historical Review* 89: 346–81.

Lenczowski, G. (1960). *Oil and the State in the Middle East.* Ithaca, NY: Cornell University Press.

———. (1975). "The Oil-Producing Countries." *Daedalus* 104 (4): 59–72.

Lenin, V. I. (1970 [1916]). *Imperialism, the Highest Stage of Capitalism.* Peking: Foreign Languages Press.

———. (1970). *The Proletarian Revolution and the Renegade Kautsky.* Peking: Foreign Language Press.

Leverett, F. and H. M. Leverett (2012). "The Mad Mullah Myth." *Harper's* November: 53–55.

Levin, Murray B. (1971). *Political Hysteria in America: The Democratic Capacity for Repression.* New York: Basic books.

Loderer, C. (1985). "A Test of the OPEC Cartel Hypothesis: 1974–1983." *Journal of Finance* 40 (3): 991–1006.

Longrigg, S. H. (1961). *Oil in the Middle East.* New York: Oxford University Press.

Lowinger, T. C. and R. Ram (1984). "Product Value as a Determinant of OPEC's Official Crude Oil Prices: Additional Evidence." *The Review of Economics and Statistics* 66 (4): 691–95.

Luxemburg, R. (1951[1913]). *The Accumulation of Capital.* New York: Routledge & Kegan Paul.

Mabro, Robert (1975). "Can OPEC Hold the Line?" *Middle East Economic Survey* Supplement, 18: 19.

———. (1998). "OPEC Behavior 1960–1998: A Review of the Literature." *Journal of Energy Literature* 4 (1): 3–27.

MacAvoy, P. W. (1982). *Crude Oil Prices As Determined by OPEC and Market Fundamentals.* Cambridge, MA: Ballinger.

Machlup, F. (1949). *The Basing-Point System.* Philadelphia, PA: Blackstone.

MacKinlay, A. Craig (1997). "Event Studies in Economics and Finance." *Journal of Economic Literature* 35 (1): 13–39.

Malthus, T. R. (1815). *An Enquiry into the Nature and Progress of Rent.* Reprints of Economic Classics. New York: Greenwood Press.

Mandel, E. (1981). "Introduction." *Marx's Capital,* vol. 3. London: Penguin.

Mandel, E. and A. Freeman, eds. (1984). *Ricardo, Marx, Sraffa.* London: Verso.

Marks, A. (2012). "Can Natural Gas Save America?" *Christian Science Monitor Weekly* 104 (22), April 23: 26–31.

Marshall, A. (1893). "On Rent." *Economic Journal* 3.

———. (1961). *Principles of Economics*. 8th ed. New York: Macmillan.

Marx, K. (1968). *Theories of Surplus Value*, vol. 2. Moscow: Progress Publishers.

Marx, K. (1969). *The Poverty of Philosophy*. New York: International Publishers.

———. (1970). *A Contribution to the Critique of Political Economy*. Moscow: Progress Publishers.

———. (1973a). *Grundrisse: Foundations of the Critique of Political Economy*. London: Penguin; New York: Vintage.

———. (1973b). *The Poverty of Philosophy*. Moscow: Progress Publishers.

———. (1977). *Capital*, vol. 1. New York: Vintage Edition.

———. (1981). *Capital*, vol. 3. London: Penguin.

Marx, K. and F. Engels (1975). *Selected Correspondence*. Moscow. Progress Publishers.

Massarrat, M. (1980). "The Energy Crisis: The Struggle for Redistribution of Surplus Profit from Oil." In P. Nore and T. Turner (eds.), *Oil and Class Struggle*, 26–68. London: Zed.

McDonald, Stephen L. (1971). *Petroleum Conservation in the United States: An Economic Analysis*. Baltimore, MD: Johns Hopkins University Press.

McKie, James (1975). "The United States." *Daedalus* 104 (4): 73–90.

McNulty, P. (1967). "Economic Theory and the Meaning of Competition." *Quarterly Journal of Economics* 82: 639–56.

Mead, W. J. (1979). "An Economic Analysis of Crude Oil Price Behavior in the 1970s." *Journal of Energy and Development* 5: 212–28.

Means, G. (1972). "The Administered Price Thesis Confirmed." *American Economic Review* June: 292–306.

Mearsheimer, J. J. and S. M. Walt (2007). *The Israel Lobby and US Foreign Policy*. New York: Farrar, Straus and Giroux.

Mikdashi Z. (1966). *A Financial Analysis of Middle Eastern Oil Concessions: 1901–65*. New York: Praeger Publishers.

———. (1972). *The Community of Oil Exporting Countries*. Ithaca, NY: Cornell University Press.

Milios, J. and D. Sotiropoulos (2009). *Rethinking Imperialism*. New York: Palgrave

Miller, E. (1973). "Some Implications of Land Ownership Pattern for Petroleum Policy." *Land Economics* 49 (4): 414–23.

Minard, L. (1980). "On the Spot." *Forbes*, January: 29–30.

Mommer, B. (2002). *Global Oil and the Nation State*. Oxford: Oxford University Press.

Moudud, Jamee, Cyrus Bina, and Patrick Mason, eds. (2013). *Alternative Theories of Competition: Challenges to the Orthodoxy*. New York and London: Routledge.

Murray, R. (1977). "Value and Theory of Rent" Part 1. *Capital and Class* 3: 100–122.

———. (1978). "Value and Theory of Rent" Part 2. *Capital and Class* 4: 11–33.

The National Security Archive (2003). "Shaking Hands with Saddam Hussein: The U.S. Tilts Toward Iraq, 1980–1984." In Joyce Battle (ed.), *National Security Archive Electronic Briefing Book* 82, February 25: http://www.gwu.edu/~nsarchiv/NSAEBB/NSAEBB82/ index.htm.

———. ed. (2003). *National Security Archive Electronic Briefing Book* 82, February 25: http://www.gwu.edu/~nsarchiv/NSAEBB/ NSAEBB82/ index.htm.

Negri, A. and M. Hardt (2000). *Empire.* Cambridge, MA: Harvard University Press.

Ng, Y. K. (1983). "Rents and Pecuniary Externalities in Cost-Benefit Analysis: Comment." *American Economic Review* 73 (5).

Nitzan, J. and S. Bichler (1995). "Bringing Capital Accumulation Back In: The Weapon-dollar-Petrodollar Coalition-Military Contractors, Oil Companies and Middle East 'Energy Conflicts'." *Review of International Political Economy* 2 (3): 446–515.

Nordhaus, William D. (2002). "Iraq: The Economic Consequences of War." *New York Review of Books* 49 (19), December 5.

Nore, P. (1980). "Oil and the State: A Study of Nationalization in the Oil Industry." In P. Nore and T. Turner (eds.) *Oil and Class Struggle*, 69–88. London: Zed.

Nwoke, C. (1987). *Third World Minerals and Global Pricing.* London: Zed.

Nye, Joseph S. (1991). *Bound to Lead: The Changing Nature of American Power.* New York: Basic Books.

———. (2004). *Soft Power: The Means to Success in World Politics.* New York: Public Affairs.

O'Conner, H. (1955). *The Empire of Oil.* New York: Monthly Review Press.

Organization of Petroleum Exporting Countries. *OPEC Conference.* Vienna, Austria, various press releases.

Painter, David S. (1986). *Oil and the American Century.* Baltimore, MD: Johns Hopkins University Press.

Pakravan, K. (1976). "The Theory of Exhaustible Resources and Market Organization, with an Application to Oil and OPEC." Unpublished PhD Dissertation, University of Chicago.

Palloix, C. (1975). "The Internationalization of Capital and the Circuit of Social Capital." in H. Radice (ed.), *International Firms and Modern Imperialism.* London: Penguin.

———. (1977). "The Self-Expansion of Capital on a World Scale." *Review of Radical Political Economics* 9 (2): 1–28.

Panitch, L. and S. Gindin (2003). "Global Capitalism and American Empire." In Leo Panitch and Colin Leys (eds.), *Socialist Register 2004.* London: Merlin: 1–42.

Panitch, L. and Colin Leys, eds. (2003). *Socialist Register 2004.* London: Merlin Press.

Pappe, Ilan (2006). *The Ethnic Cleansing of Palestine*. Oxford: Oneworld Publications.
Penrose, E. (1968). *The Large International Firm in Developing Countries: The International Petroleum Industry*. London: Allen and Unwin.
———. (1975). "The Development of Crisis." *Daedalus* 104 (4): 39–57.
Peterson, S. (2012a). "How Iran would Fight US." *Christian Science Monitor Weekly* 104 (11), February 6: 8–9.
———. (2012b). "What if Iran Gets the Bomb?" *Christian Science Monitor Weekly* 104 (14), February 27: 26–31.
Pindyck, Robert (1978). "Gains to Producers from the Cartelization of Exhaustible Resources." *Review of Economics and Statistics* 60 (2): 238–51.
———. (1979). "Some Long-Term Problems in OPEC Oil Pricing." *Journal of Energy and Development* 4 (2): 259–72.
Poulantzas, N. (1975). *Classes in Contemporary Capitalism*. London: Verso.
Prast, W. G. and H. L. Lax (1983). *Oil Futures Markets: An Introduction*. Lexington, MA: Lexington Books.
Prebisch, Raúl (1950). *The Economic Development of Latin America and Its Principal Problems*. New York: United Nations.
Ricardo, D. (1976). *The Principles of Political Economy and Taxation*. 3rd ed. London: J. M. Dent & Sons Ltd.
Rifai, T. (1974). *The Pricing of Crude Oil*. New York: Praeger Publishers.
Risen, James (2006). *State of War*. New York: Free Press.
Robinson, Joan (1969[1933]). *The Economics of Imperfect Competition*. London: Macmillan.
Rosdolsky, R. (1977). *The Making of Marx's 'Capital'*. London: Pluto Press.
Rouhani, F. (1971). *A History of OPEC*. New York: Praeger Publishers.
Saad-Filho, A. (1993). "A Note on Marx's Analysis of the Composition of Capital." *Capital and Class* 50: 127–46.
Salant, S. W. (1982). *Imperfect Competition in the World Oil Market: A Computerized Nash Cournot Model*. Lexington, MA: Lexington Books.
Salehi-Isfahani, D. (1987). "Testing OPEC Behavior: Further Results." Department of Economics, VPI and SU, Working Paper # 87–01–02.
———. (1995). "Models of the Oil Market Revisited." *The Journal of Energy Literature* 1 (1): 3–21.
Sampson, A. (1975). *The Seven Sisters: The Great Oil Companies and the World They Made*. London: Hodder and Stoughton.
Sand, Shlomo (2012). *The Invention of the Land of Israel*. London and New York: Verso.
Sassoon, A. Showstack (1982). "Passive Revolution and the Politics of Reform." In Sassoon (ed.), *Approaches to Gramsci*. London: Writers and Readers.
Schell, J. (2012). "Thinking the Unthinkable" *The Nation* April 23: 20–26.
Schlesinger, James (1991). "Interview: Will War Yield Oil Security?" *Challenge* March/ April.
Schumpeter, J. (1928). "The Instability of Capitalism." *The Economic Journal* 38 (151): 361–86.

———. (1942). *Capitalism, Socialism and Democracy.* New York: Harper & Row.

———. (1954). *History of Economic Analysis.* New York: Oxford University Press.

Semmler, W. (1982). "Competition, Monopoly, and Differentials of Profit Rates: Theoretical Consideration and Empirical Evidence." *Review of Radical Political Economics* 13 (4).

———. (1984). *Competition, Monopoly, and Differential Profit Rates.* New York: Columbia University Press.

Shaikh, A. (1977). "Marx's Theory of Value and the Transformation Problem." In J. Schwartz (ed.), *The Subtle Anatomy of Capitalism*, 106–37. Santa Monica, CA: Goodyear.

———. (1978). "An Introduction to the History of Crisis Theories." In The Crisis Reader Editorial Collective, *U.S. Capitalism in Crisis.* URPE: New York.

———. (1980). "Marxian Competition versus Perfect Competition: Further Comments on the So-Called Choice of Techniques." *Cambridge Journal of Economics* 4 (1): 75–83.

———. (1981). "Differential Rent." *MIMEO.* New York: New School for Social Research.

———. (1982). "Neo-Ricardian Economics: A Wealth of Algebra, A Poverty of Theory." *Review of Radical Political Economics* 14 (2): 67–74.

———. (1984). "The Transformation from Marx to Sraffa." In E. Mandel and A. Freeman (eds.), *Ricardo, Marx, Sraffa*, 43–84. London: Verso.

———. (1990). "Organic Composition of Capital." In J. Eatwell, M. Milgate, and P. Newman (eds.), *The New Palgrave—Marxian Economics*, 304–09. New York: W.W. Norton.

Shanker, T. and E. Schmitt (2004). "Pentagon Weighs Use of Deception in a Broad Arena." *New York Times*, December 13: 1A, 12A.

Sherif, Regina S. ed. (1988). *United Nations Resolutions on Palestine and the Arab-Israeli Conflict, 1975–1981.* vol. II Washington, DC: Institute for Palestine Studies.

Simpson, Michael ed. (1988). *United Nations Resolutions on Palestine and the Arab-Israeli Conflict, 1982–1986.* vol. III Washington, DC: Institute for Palestine Studies.

Smith, A. (1977 [1776]). *The Wealth of Nations.* New York: Penguin.

Smith, James L. (2003). "Inscrutable OPEC? Behavioral Tests of the Cartel Hypothesis." Department of Finance. Southern Methodist University, Dallas, TX, mimeograph.

Smithies, A. (1949). "Economic Consequences of the Basing-Point Decisions." *Harvard Law Review* 62 (2): 308–18.

Sraffa, P. (1960). *Production of Commodities by Means of Commodities.* Cambridge: Cambridge University Press.

Steedman, I., P. Sweezy, et al. (1981). *The Value Controversy.* London: Verso.

Steel, Ronald (1967). *Pax Americana.* New York: Viking Press.

Steindl, J. (1952). *Maturity and Stagnation in American Capitalism.* New York: Monthly Review Press.

Stiglitz, Joseph E. and Linda Bilmes (2008). *The Three Trillion Dollar War: The True Cost of the Iraq Conflict.* New York: W. W. Norton.

Stokes, D. and S. Raphael (2010). *Global Energy Security and American Hegemony.* Baltimore, MD: John Hopkins University Press.

Stock, F. (1980). "Transition in Energy Resources." *Petroleum Economist,* May.

Stocking, G. W. (1950). "The Economics of Basing-Point System." *Law and Contemporary Problems* 15 (2): 159–80.

Stocking, G. W. and M. W. Watkins (1948). *Cartels or Competition.* New York: Twentieth Century Fund.

Stork, J. (1975). *Middle East Oil and the Energy Crisis.* New York: Monthly Review.

Suskind, Ron (2006). *The One Percent Doctrine.* New York: Simon & Schuster.

Sweezy, Paul (1942). *The Theory of Capitalist Development.* New York: Monthly review Press.

Symposium on Exhaustible Resources (1974). *Review of Economic Studies.*

Tanzer, M. (1969). *The Political Economy of International Oil and the Underdeveloped Countries.* Boston, MA: Beacon Press.

———. (1974). *The Energy Crisis: World Struggle for Power and Wealth.* New York: Monthly Review Press.

———. (1980). *The Race for Resources: Continuing Struggles over Minerals and Fuels.* New York: Monthly Review Press.

Tariki, A. (1963). "Towards Better Cooperation between Oil Producing and Oil Consuming Countries." *Petroleum Intelligence Weekly* November 11 (Special Supplement).

Teece, D. J. (1982). "OPEC Behavior: An Alternative View." In J. M. Griffin and D. J. Teece (eds.), *OPEC Behavior and World Oil Prices.* London: Allen and Unwin.

Terzian, P. (1985). *OPEC: An Inside Story.* London: Zed Press.

Thomas, J. (1992). "Cartel Stability in an Exhaustible Resource Model." *Economica* New Series, 59: 235, 279–93.

Thomas, Peter D. (2009). *The Gramscian Moment.* Chicago, IL: Haymarket Books.

Thomson, R. (1990). "Primitive Capitalist Accumulation." In J. Eatwell, M. Milgate, and P. Newman (eds.), *The New Palgrave—Marxian Economics,* 313–20. New York: W. W. Norton.

Thompson, E. P. (1978). *The Poverty of Theory and Other Essays.* New York: Monthly Review Press.

Tomeh, George J. ed. (1975). *United Nations Resolutions on Palestine and the Arab-Israeli Conflict, 1947–1974.* vol. I Washington, DC: Institute for Palestine Studies.

Torrens, R. (1814). *An Essay on the External Corn Trade.* 1st ed. Reprints of Economic Classics (1972). New York: Augustus M. Kelley Publishers.

Tsurumi, Yoshi (1975). "Japan." *Daedalus* 104 (4): 113–27.

UN General Assembly—GA/11317 (2012). "General Assembly Votes Overwhelmingly to Accord Palestine." *United Nations*. New York: Department of Public Information, November 29: http://www.un.org/News/Press/docs//2012/ga11317.doc.htm.

US Congress, House Committee on Foreign Relations (1990). Hearing before the Subcommittee on Europe and the Middle East, *United States-Iraq Relations*. Washington, DC: US Government Printing Office, April 26.

US Congress, Senate Committee on Energy and Natural Resources (1977). *Petroleum Industry Involvement in Alternative Sources of Energy*. Washington, DC: US Government Printing Office.

US Congress, Senate Committee on Foreign Relations (1987). *War in the Persian Gulf: The U.S. Takes Sides*. Washington, DC: US Government Printing Office, November.

US Department of Energy (1978a). *Depth and Producing Rate Classification of Domestic Oil Reservoirs (1974)*. Washington, DC: US Government Printing Office.

———. (1978b). *An Analysis of the Productivity of Domestic Petroleum Exploration Activities*. Washington, DC: US Government Printing Office.

US Department of the Interior (1967). "Depth and Producing Rate Classification of Oil Reservoirs in the 14 Principal Oil-Producing States." *Bureau of Mines Information Circular (8362)*. Washington, DC: US Government Printing Office.

———. (1976). "Depth and Producing Rate Classification of Petroleum Reservoirs in the United States." *Bureau of Mines Circular (8675A)*. Washington, DC: US Government Printing Office.

US Government Printing Office (1980). *Coal Data Book*. Washington, DC.

US-UK Memorandum (1964). "Energy Diplomacy and Global Issues." In *Foreign Relations of the United States 1964–1968* 34 (1999): 317–20. Washington, DC: Government Printing Office.

Verleger, Jr. P. K. (1982a). "The Determinants of Official OPEC Crude Prices." *The Review of Economics and Statistics* 64 (2): 177–83.

———. (1982b). *Oil Markets in Turmoil*. Cambridge, MA: Ballinger.

Vernon, R. (1975a). "An Interpretation," *Daedalus* 104 (4): 1–14.

———. ed. (1975b). "The Oil Crisis in Perspective." *Daedalus*, Fall.

Vidal, Gore (2002). "Blood for Oil." *The Nation* October 28.

Walden, J. L. (1962). "The International Petroleum Cartel in Iran—Private Power and Public Interest." *Journal of Public Law* (11): 3–60.

Wallerstein, I. (1979). *The Capitalist World-Economy*. Cambridge: Cambridge University Press.

———. (2002). "The Eagle Has Crash Landed." *Foreign Policy*, July/August: http://www.foreignpolicy.com/issue_julyaug_2002/ wallerstein.ht

Weeks, J. (1981a). *Capital and Exploitation*. Princeton, NJ: Princeton University Press.

———. (1981b). "The Differences between Materialist Theory and Dependency Theory and Why They Matter." *Latin American Perspectives* 8 (3–4): 118–23.

Weeks, J. (1983). "On the Issue of Capitalist Circulation and the Concepts Appropriate to Its Analysis." *Science & Society* 47 (2): 214–25.
Weeks, J. (2011). *Capital, Exploitation and Economic Crisis.* London: Routledge.
———. (2013). "The Fallacy of Competition." In J. Moudud, C. Bina, and P. Mason (eds.), *Alternative Theories of Competition: Challenges to the Orthodoxy*, 13–26. London: Routledge.
Wessel, R. H. (1967). "A Note on Economic Rent." *American Economic Review* 57.
West, Sir Edward (1815). "Essay on the Application of Capital to Land." Reprinted in J. H. Hollander (ed.), *A Reprint of Economic Tracts* (1903). Baltimore, MD: Johns Hopkins Press.
White House (2001). *Report of the National Energy Policy Development Group.* Washington, DC: Government Printing Office, May.
———. (2002). *The National Security Strategy of the United States of America.* Washington, DC: Government Printing Office, September 17.
Willoughby, John (1986). *Capitalist Imperialism: Crisis and the State.* New York: Harwood Academic Publishers.
Wilson, T. (1979). "The Price of Oil: A Case of Negative Marginal Revenue." *Journal of Industrial Economics* 27 (4).
Woodward, Bob (2004). *Plan of Attack.* New York: Simon & Schuster.
Wyant, Frank R. (1979). *The United States, OPEC, and Multinational Oil.* Lexington, MA: D.C. Heath.
Yamani, A. Z. (1969). "Participation Versus Nationalization: A Better Means to Survive." *Middle East Economic Survey* 12: June 13.
Yergin, Daniel (1991). *The Prize: The Epic Quest for Oil, Money, and Power.* New York: Simon & Schuster.
———. (2011). *The Quest: Energy, Security, and the Remaking of the Modern World.* New York: Penguin.
Zangeneh, H., ed. (1994). *Islam, Iran, and World Stability.* New York: St. Martin's.
Zoninsein, Jonas (1990). *Monopoly Capital Theory: Hilferding and Twentieth-Century Capitalism.* New York: Greenwood Press.

# Index

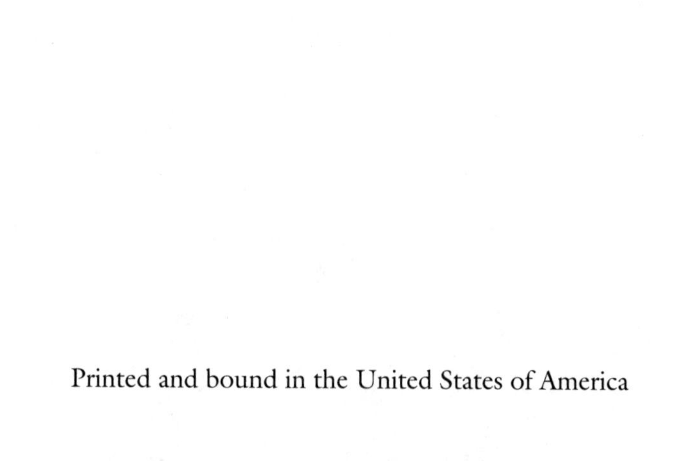

Printed and bound in the United States of America